KU-737-294

Series Editor's Foreword

Modern organic chemistry owes its existence to the fascination chemists have had, and continue to have, with natural products: their isolation, their synthesis, and above all the elucidation of their biosynthesis and the control elements therein have motivated many generations.

Oxford Chemistry Primers have been designed to provide concise introductions relevant to all students of chemistry and contain only the essential material that would be covered in an 8–10 lecture course. In this twentieth Primer on chemical aspects of biosynthesis, John Mann provides an enthusiastic account of this central area of organic chemistry that is guaranteed to stimulate apprentice and master chemist alike.

Stephen G. Davies
The Dyson Perrins Laboratory, University of Oxford

Preface

Human cultures have always been fascinated by natural products. They extracted them from plants, animals, and lesser organisms, and then used them in crude form for the treatment of disease, as poisons, and as euphoriants and stimulants. During the nineteenth century and most of the present one, structural and synthetic studies have been the main preoccupation of chemists, and thousands of natural products have been identified and synthesized.

The availability of the radioisotopes ^{14}C and tritium at the end of World War II allowed biosynthetic investigations to commence, and during the past 50 years the elucidation of the pathways to the major classes of natural products—polyketides, shikimate metabolites, isoprenoids, and alkaloids—has been achieved. This book describes the most important results of these endeavours with an emphasis on the chemistry involved in the biosynthetic pathways. The pharmacological, toxicological, and ecological significance of the compounds is also highlighted, since these properties underpin the human use of natural products and their involvement in chemical communication between organisms.

The book is based upon a course of 15 lectures given to the Honours Chemists at the University of Western Australia and subsequently given (in an abbreviated form) to the students at Reading. I thank Dieter Wege and Mel Sargent for inviting me to UWA as part of their Visiting Lecturer Programme, and for making the visit both memorable and productive. Two superb biosynthetic investigators, Jim Hanson and Tom Simpson, read all or part (respectively) of the manuscript, and their comments and criticisms were invaluable. But my main thanks must be reserved for Mike Harling who drew almost all of the structures. Without his expert assistance the project would have taken twice as long to complete. Finally, I should like to dedicate the book to Derek Banthorpe, my Ph.D. supervisor, who excited my interest in natural products more than a quarter century ago.

J.M.

Reading
July 1994

Contents

Chemical Aspects of Biosynthesis

John Mann

Department of Chemistry,
University of Reading

LIVERPOOL JOHN MOORES UNIVERSITY
LEARNING SERVICES

Accession No
BM 267393 K

Class No 572.45 MAN

Aldham Robarts	Byrom Street	✓
0151-231-3701	0151-231-2329	
I M Marsh	Trueman Street	
0151-231-5216	0151-231-4022	

LIVERPOOL
JOHN MOORES UNIVERSITY
AVRIL ROBARTS LRC
TITHEBARN STREET
LIVERPOOL L2 2ER
TEL. 0151 231 4022

OXFORD
UNIVERSITY PRESS

LIVERPOOL JMU LIBRARY

3 1111 00880 3049

OXFORD

UNIVERSITY PRESS

Great Clarendon Street, Oxford OX2 6DP

Oxford University Press is a department of the University of Oxford.
It furthers the University's objective of excellence in research, scholarship,
and education by publishing worldwide in

Oxford New York

Athens Auckland Bangkok Bogotá Buenos Aires Calcutta
Cape Town Chennai Dar es Salaam Delhi Florence Hong Kong Istanbul
Karachi Kuala Lumpur Madrid Melbourne Mexico City Mumbai
Nairobi Paris São Paulo Singapore Taipei Tokyo Toronto Warsaw

with associated companies in Berlin Ibadan

Oxford is a registered trade mark of Oxford University Press
in the UK and in certain other countries

Published in the United States
by Oxford University Press Inc., New York

© John Mann, 1994

The moral rights of the author have been asserted
Database right Oxford University Press (maker)

First published 1994
Reprinted 1996, 1999

All rights reserved. No part of this publication may be reproduced,
stored in a retrieval system, or transmitted, in any form or by any means,
without the prior permission in writing of Oxford University Press,
or as expressly permitted by law, or under terms agreed with the appropriate
reprographics rights organization. Enquiries concerning reproduction
outside the scope of the above should be sent to the Rights Department,
Oxford University Press, at the address above

You must not circulate this book in any other binding or cover
and you must impose this same condition on any acquirer

A catalogue record for this book is available from the British Library

Library of Congress Cataloging in Publication Data
Mann, J.
Chemical aspects of biosynthesis/John Mann.—1st ed.
p. cm.—(Oxford chemistry primers; 20)
Includes bibliographical references.
I. Biosynthesis. 2. Metabolism, Secondary. 3. Metabolites.
I. Title. II. Series.
QP517.B57M36 1994 574.19'29—dc20 94-10981

ISBN 0 19 855677 2 (Hbk.)
ISBN 0 19 855676 4 (Pbk.)

Printed in Great Britain on acid-free paper by
The Bath Press.

1 Introduction

'The Indians killed another companion of ours . . . and in truth, the arrow did not penetrate half a finger, but, as it had poison on it, he gave up his soul to our Lord.'

Thus wrote the Spaniard Francisco de Orellano, who accompanied the conquistadores on their murderous campaign through South America in the sixteen century. The poison in question was almost certainly curare, a collective name for extracts of the plant *Chondodendron tomentosum* and various *Strychnos species*. The major active ingredient is the natural product tubocurarine and this has potent activity as a neuromuscular blocking agent. It functions by binding to a receptor for the neurotransmitter acetylcholine at the point where nerve ending impinges on a muscle endplate. By denying access to acetylcholine, neurotransmission is prevented, and there is a resultant muscle paralysis.

Other natural products are shown in Fig. 1.1 and, like tubocurarine, these molecules are both structurally and pharmacologically interesting. Most of the compounds illustrated have been used in crude form for centuries, but were first isolated in pure form in the nineteenth century. Structure elucidation was often painfully slow, and necessitated lengthy chemical transformations and degradations. The advent of modern chromatographic and spectroscopic methods now usually allows a rapid elucidation of structures of newly isolated compounds. In consequence, this book will be concerned primarily with the ways in which natural products are assembled, and with their pharmacological and ecological significance.

MEDICINES

Ricinine
(castor oil: purgative)

Salicin
(from Willow bark:
used by 'rural folk' for fevers;
aspirin is a synthetic analogue)

Ephedrine
(respiratory ailments)

Quinine from Cinchona bark

NARCOTICS AND HALLUCINOGENS

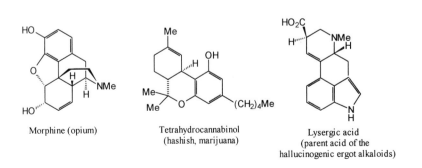

Morphine (opium)

Tetrahydrocannabinol
(hashish, marijuana)

Lysergic acid
(parent acid of the
hallucinogenic ergot alkaloids)

Fig. 1.1

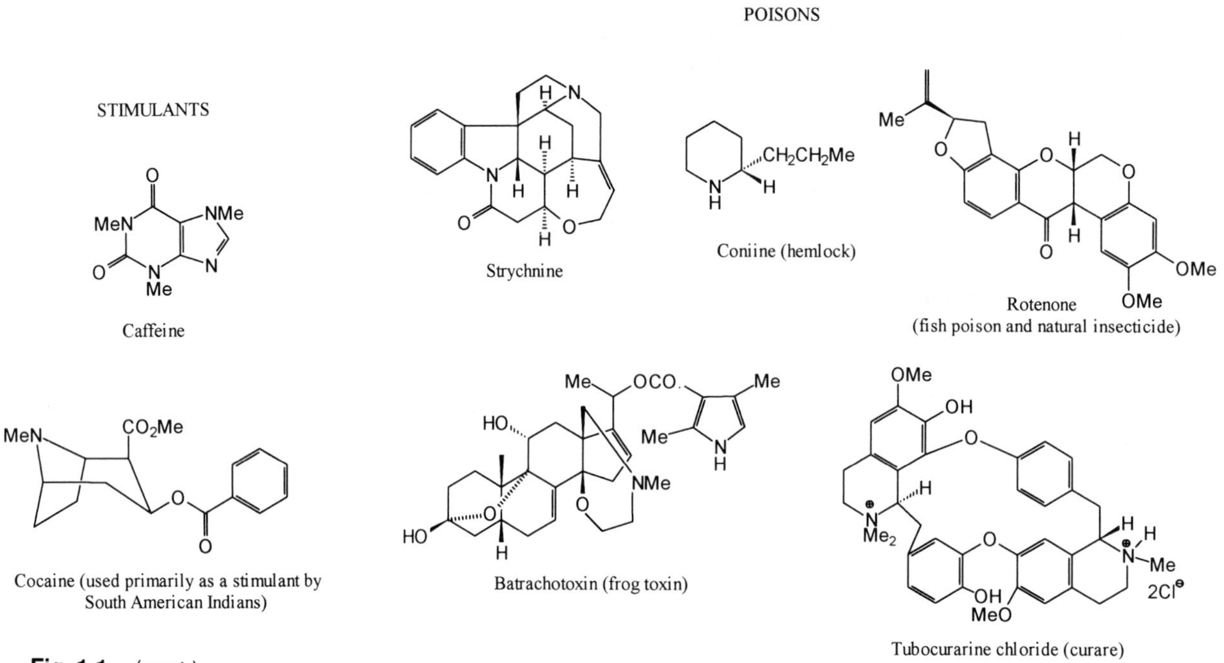

POISONS

STIMULANTS

Strychnine

Coniine (hemlock)

Rotenone
(fish poison and natural insecticide)

Caffeine

Cocaine (used primarily as a stimulant by
South American Indians)

Batrachotoxin (frog toxin)

Tubocurarine chloride (curare)

Fig. 1.1 (cont.)

1.1 Primary and secondary metabolites

There are two major classes of natural products, primary and secondary metabolites. The ones shown in Fig. 1.1 are usually called 'secondary metabolites'. They are compounds that are unique to a particular species or of relatively limited occurrence—morphine, which comes from just two species of poppy, *Papaver somniferum* and *P. setigerum*, and the antimalarial agent quinine, obtained solely from the *Cinchona* tree, nicely exemplify this feature. One could also add that secondary metabolites usually have no proven effect on the organisms producing them, though they may have a deterrent effect on would-be predators or competitors.

Secondary metabolites are all produced from a relatively small number of key intermediates. These arise from the primary metabolic pathways that are shared by all organisms, namely the pathways for the production (anabolism) or metabolism (catabolism) of primary metabolites: carbohydrates, proteins, and nucleic acids. These primary pathways of metabolism and their links with secondary metabolism are shown in Fig. 1.2.

From this figure it will be apparent that there are several principal building blocks for the biosynthesis of secondary metabolites: acetate, mevalonate, shikimate, and the amino acids. Acetate in the form of its thioester with coenzyme A (see below) is the precursor for the production of many phenols and other aromatic species, and of fatty acid metabolites like the prostaglandins and leukotrienes. Mevalonate (itself derived from acetyl thiocoenzyme A) is the progenitor of the terpenoids and steroids; shikimic acid is used in the production of other aromatic compounds; and the amino acids are the biosynthetic precursors of the alkaloids.

Most of these key intermediates are also employed for the construction of primary metabolites like nucleosides, fatty acids, and polypeptides, and it is worth asking the question: what determines whether they act as precursors of primary or secondary metabolites? Various explanations and theories have been considered, and it is probable that the secondary metabolic pathways evolved as a means of consuming acetate, shikimate, and amino

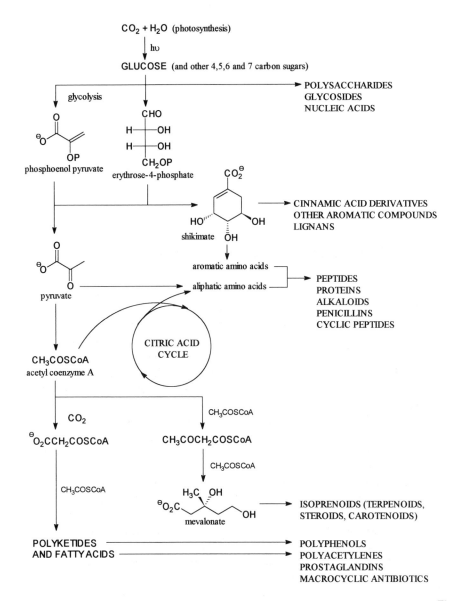

Fig. 1.2

acids that were surplus to the requirements of the primary metabolic pathways. In a sense they are thus products of 'overflow' metabolism or 'shunt metabolites'. This of course evades the questions of why the key intermediates were produced in excess and how the new chemistry evolved. Possible answers to both questions can be found if we consider the essential features of the evolutionary process.

Organisms evolve through changes (mutations) in their genetic make-up, and since one gene generally specifies one or more proteins or enzymes, concomitant changes in protein structure or enzyme activity result. The mutations may result from exposure of the organism to ionising radiation, UV irradiation, certain chemicals, or viruses. However, they arise most often as a result of 'natural chemical mistakes' in the complex processes whereby a gene is copied and ultimately acts as a blueprint for the production of proteins. New enzymes may be produced or there may simply be an elevation in the amount of an existing enzyme. Genetic changes that result in alterations to primary metabolic pathways are usually detrimental and the organism may die, though major evolutionary innovation occurs by adoption of these changes. Survival may be enhanced if an organism produces a better means of defence or new enzymes for the digestion of foods that were hitherto toxic to it. For example at some stage in their evolution, various species of South American frogs began to produce toxic alkaloids like the batrachotoxins (see Fig. 1.1). These enhanced their chances of survival (and hence their chances of passing on their genes through reproduction) because they were distasteful to predators. The koala bear, in contrast, produced the necessary digestive enzymes to cope with a diet rich in eucalyptus terpenoids like 1,8-cineole. This enabled it to spend its entire existence in the tops of eucalyptus trees thus avoiding the attentions of ground-based predators.

Many of the genetic changes probably led to an increased production of acetate, shikimate, and other key intermediates, and to new enzymes which catalysed novel transformations of these compounds. The products of this new chemistry were the secondary metabolites. If these new compounds possessed useful biological activity, such as deterrence or attraction for other members of the species (thus enhancing the chances of successful mating), the survival prospects of the organism would be increased. The genetic changes would thus be more likely to be passed on to the next generation via mating.

Secondary metabolites are thus not usually mere products of waste metabolism, and some of them are probably vital for the survival of the organism or species. They are certainly very prevalent in the lower organisms and in organisms that occupy precarious ecological niches. The higher organisms, for example man, have few secondary metabolites since their survival is now more dependent upon physical means of defence and attraction. Chemical communication has given way to the superior communication possibilities provided by a central nervous system.

The production of secondary metabolites is obviously dependent upon the interconversions catalysed by enzymes and their cofactors, and we should consider briefly the main processes of oxidation and reduction, carbon–carbon bond formation and carbon–nitrogen bond formation.

1.2 Oxidation and reduction

In the laboratory, oxidation of organic molecules is most usually effected with Cr (VI) reagents or with highly oxidised forms of manganese, osmium, or ruthenium (e.g. $KMnO_4$, OsO_4, and RuO_4). *In vivo* it is almost invariably effected by molecular oxygen or by an active oxygen species like superoxide ($O_2^{\bar{\cdot}}$) or peroxide (O_2^{2-}). Laboratory reductions are most conveniently carried out with one of the many hydride reducing agents ($LiAlH_4$, $NaBH_4$, etc.), and here the parallels with *in vivo* reductions are more apparent.

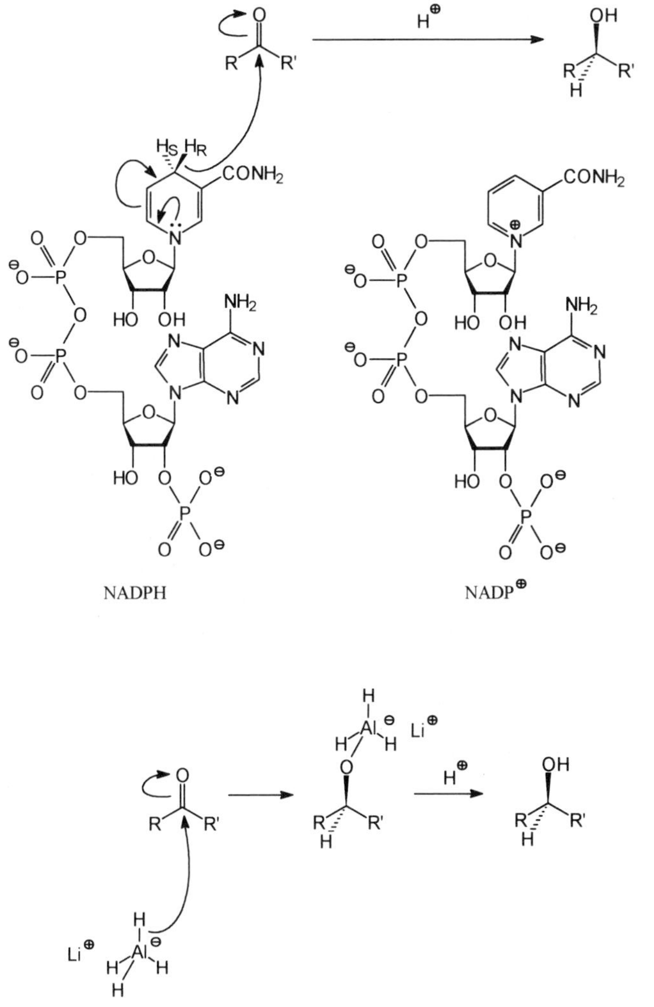

NADPH   NADP$^{\oplus}$

Fig. 1.3

Enzymes called dehydrogenases, working in consort with the cofactor nicotinamide adenine dinucleotide phosphate (NADPH), catalyse the reduction of ketones and aldehydes to alcohols, according to the mechanism shown in Fig. 1.3. One of the C-4 hydrogens of the nicotinamide ring is transferred (as hydride) to the carbonyl carbon, and the resultant oxyanion reacts with a proton supplied by the enzyme or, more probably, from water. Compare the situation that exists in a reduction involving $LiAlH_4$, where hydride is derived from the reagent, and a proton from water.

The main difference is that the two potential hydrides of NADPH are different since the 4-carbon is a prochiral centre: that is, if one of the hydrogens were to be replaced by deuterium or tritium, the carbon would become chiral. Since the cofactor and the substrate are usually held in a defined arrangement within the three-dimensional environment of the enzyme's active site, it is usually only possible for one or other of the hydrides to be transferred. The reduction thus becomes stereospecific with transfer of the H_R or H_S hydrogen to one or other face of the carbonyl compound, and a chiral reduction product is obtained.

In vivo oxidations are catalysed by four main classes of enzymes: oxidases, oxygenases, peroxidases, and catalases. The oxidases catalyse the transfer of hydrogen from the substrate to molecular oxygen, and flavin cofactors are usually involved. Unlike NADPH which can migrate from enzyme to enzyme, the flavin cofactors (FAD and $FADH_2$) are almost invariably tightly bound to one particular enzyme. The actual mechanism of hydrogen transfer is not yet known, but reoxidation of the reduced cofactor ($FADH_2$) is normally effected directly or indirectly by molecular oxidation. This probably involves one-electron transfers to oxygen to produce the superoxide radical anion (O_2^-) and thence the peroxide dianion (O_2^{2-}).

The most important oxygenases are those of the cytochrome P_{450} family. These incorporate a haem cofactor and absorb light maximally around 450 nm when complexed with CO—hence their name. The haem iron has two vacant coordination sites and these are occupied by a cysteine thiolate (from the enzyme) and oxygen or a derived species. They catalyse the reaction

$$R-H + O_2 + 2H^+ + 2e \longrightarrow ROH + H_2O$$

and are involved in the oxidation of arenes to phenols, alkenes to epoxides, alkanes to alcohols, and sulphides to sulphoxides.

There is now excellent experimental evidence for the free radical mechanism shown below:

Probably the closest *in vitro* analogy for these free radical oxidations would be the oxidation (with $KMnO_4$) of carbon side-chains attached to aryl systems, since this involves a conversion of R–H to R–OH via a free radical mechanism.

1.3   Carbon–carbon bond formation

In the laboratory, new carbon–carbon bonds are most commonly created when a carbon nucleophile (present in a Grignard reagent, organolithium reagent, or phosphorus ylid) attacks a carbon electrophile. *In vivo* this process often proceeds via thioesters of the cofactor coenzyme A, and the conversion of acetyl thiocoenzyme A into hydroxymethylglutaryl thiocoenzyme A is an example (Fig. 1.4). As will become apparent in Chapter 4, this interconversion comprises the early stages of terpenoid and steroid biosynthesis. In purely mechanistic terms, a Claisen ester condensation is followed by an aldol reaction, and the thioester linkage serves both as a convenient means of anchoring the substrates to the cofactors, and as a good leaving

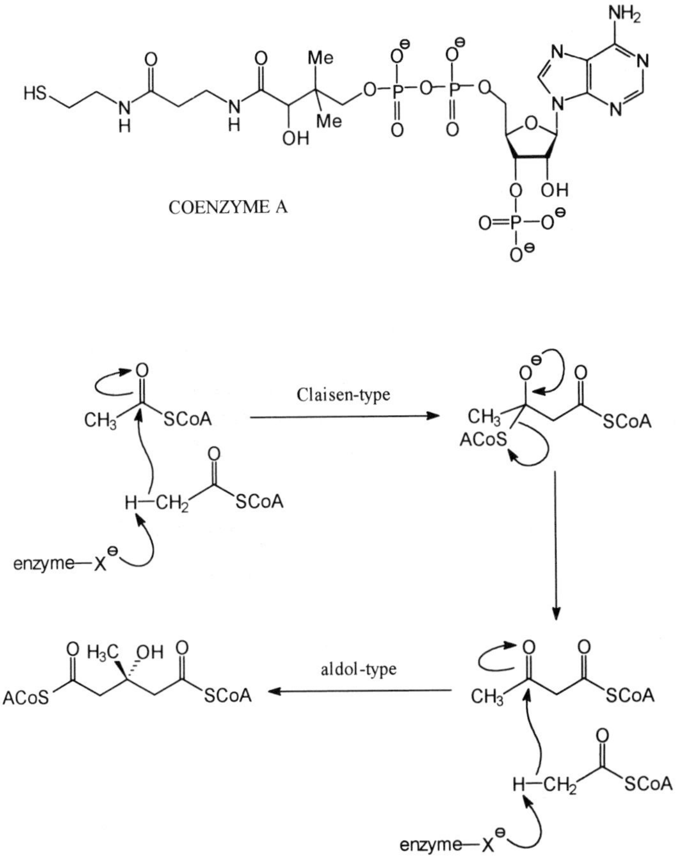

Fig. 1.4

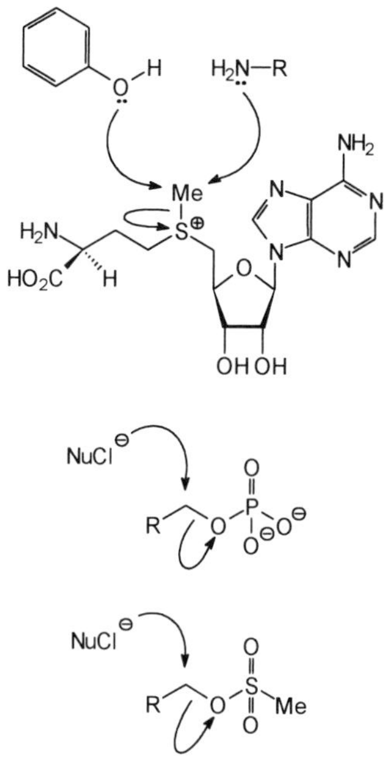

Fig. 1.5

group when participating in the Claisen condensation. Note also the use of pyrophosphate esters (shown in Fig. 1.5). The chemistry is analogous to the use of methanesulphonates or toluenesulphonates in the laboratory, and all of these groups are readily displaced by nucleophiles thus increasing the electrophilic nature of the carbon to which the groups are attached.

The cofactor *S*-adenosylmethionine serves as a source of a one-carbon electrophile (Me^+), and is involved in carbon–carbon bond formation as well as in the formation of OMe and NMe groups, as shown in Fig. 1.5. Bonds between nitrogen and other alkyl groups are usually formed by reaction of amines and aldehydes or ketones followed by reduction of the resultant condensation products (Schiff's bases). This chemistry is particularly important in the biosynthesis of alkaloids and will be covered in more detail in Chapter 6.

1.4 Elucidation of biosynthetic pathways

After structure elucidation, it is usually possible to propose a biogenetic hypothesis for the formation of the secondary metabolite from one of the key building blocks—acetate, mevalonate, shikimate, or amino acids. Actual pathways of biosynthesis can only be considered delineated once each individual intermediate (and enzyme) on these pathways has been identified.

In order to investigate the reality of these biogenetic hypotheses, it is necessary to carry out experiments in which isotopically enriched precursors are administered to the plant or microorganism producing the secondary metabolite. The metabolite is isolated, purified, and analysed for isotopic content, and if incorporation of isotope has occurred in the predicted fashion, cautious acceptance of the precursor/metabolite relationship can be assumed. It may also be possible to isolate and identify proposed intermediates which have incorporated isotope, and these may then be used as advanced intermediates in further incorporation studies:

$$A \longrightarrow B \longrightarrow C \longrightarrow \longrightarrow \longrightarrow X \longrightarrow Y \longrightarrow Z$$

A is a small, precursor species (e.g. acetate), whilst B, C, X, and Y are intermediates *en route* to the metabolite Z.

The use of mutant strains of organisms may also provide useful information, since they will almost inevitably contain aberrant enzymes or altered levels of normal enzymes. Thus, for example, the enzyme catalysing the conversion of B into C may be absent or defective, and metabolite B should then accumulate, though it may also be diverted into new or separate pathways.

Two difficulties confront the investigator in such studies:

(1) incorporation of a sufficiently high percentage of labelled precursor to render the results meaningful; and

(2) the need to analyze (and perhaps degrade) the labelled metabolite in order to establish which atoms have been isotopically enriched.

Administration of precursor

Isotopically labelled precursor may be administered to intact organisms or to cell-free extracts of them. In general, intact plants incorporate label rather poorly (often as little as 10^{-2} to 10^{-4} per cent of the total administered), usually because added precursor cannot penetrate to the site of biosynthesis or is metabolised *en route*. Better results can be obtained by using cell-free systems or tissue cultures derived from plants. The latter are usually obtained from small pieces of plant tissue grown in such a way that they produce undifferentiated callus cells, which may retain the capability to produce secondary metabolites, but have lost the capacity to produce other compounds. One eventual commercial aim of this technique is to provide cell lines that produce large amounts of, for example, specific medicinally useful alkaloids or selected terpenoids for the perfumery industry. Better incorporations of isotope are achieved with bacterial and fungal cultures, and much of the work in the past twenty years has been carried out on mould metabolites.

Examination of labelled metabolites

Once a pure sample of isotopically enriched metabolite or biosynthetic intermediate has been obtained, it must be examined to determine the location of the enriched atoms. The commonly used radioactive isotopes are tritium (^3H), a β-emitter with a half-life of 12.1 years, and carbon-14 (^{14}C), a β-emitter with a half-life of 5640 years. The radiolabelled metabolite must thus be degraded chemically to provide information about the centres of enrichment.

All of the common degradative methods have been employed: ozonolysis, decarboxylation, Hofmann degradation, etc. However, since each degradative step reduces the amount of metabolite available for further degradation, it is usually impossible to obtain a complete labelling pattern for the molecule. The use of the non-radioactive isotope carbon-13 (^{13}C), and to a lesser extent the isotopes ^2H, ^{15}N, and ^{18}O, have thus revolutionised the study of biosynthetic pathways since these isotopes have nuclear spins $I = n/2$ and are thus detectable using NMR techniques.

Since the natural abundance of the ^{13}C isotope is only 1.1 per cent, it is now only necessary to assign the natural abundance ^{13}C spectrum of a metabolite, and then compare this with the spectrum exhibited by the same compound isolated following an incorporation experiment using a ^{13}C-enriched precursor, to determine the carbon centres that originate from the precursor molecule. In addition, the use of precursors doubly labelled with ^{13}C and one of the other isotopes results in ^{13}C spectra in which some of the signals exhibit isotope-induced shifts. This provides extra structural and biosynthetic information. These techniques will be described more fully in Chapter 3.

2　Fatty acids and derivatives

$$H_2C-O\overset{O}{\overset{\|}{C}}-(CH_2)_{16}CH_3$$
$$H-\overset{|}{C}-O\overset{O}{\overset{\|}{C}}-(CH_2)_{16}CH_3$$
$$H_2C-O\overset{O}{\overset{\|}{C}}-(CH_2)_{16}CH_3$$

2.1

$$CH_3(CH_2)_7 \diagdown \diagup (CH_2)_7CO_2H$$
$$\underset{H}{\ } \quad \underset{H}{\ }$$

2.2

2.3

The most abundant naturally occurring saturated fatty acids have the composition $CH_3 (CH_2) n CO_2H$ (n = 12, 14, 16, and 18), and are vital constituents of natural waxes, seed oils, and glycerides, e.g. **2.1**. The latter are most commonly encountered as constituents of cell membranes, and have the typical polar head groups and lipophilic tails that are essential for the structural integrity and selective permeability of these membranes (Figure 2.1). In addition to this structural role, the glycerides (or fats) provide an energy store, since they can be broken down to acetate which can then be used for biosynthesis or fed into the citric acid cycle for the construction of other small building blocks.

The (poly) unsaturated fatty acids, e.g. oleic acid, **2.2**, and arachidonic acid (eicosatetra-5, 8, 11, 14-enoic acid), **2.3**, are also found in cell membranes, but in addition act as precursors of biologically active metabolites. For example, oleic acid is the biosynthetic precursor of polyacetylenes like crepenynic acid, **2.4**, and wyerone, **2.5**. The former is produced by many species of Compositae (daisy) and has anti-microbial activity; the latter is produced by the broad bean in response to stress, and has antifungal activity. Arachidonic acid is the biosynthetic precursor of a large family of metabolites known as eicosanoids, which include the pharmacologically important prostaglandins and leukotrienes.

$$CH_3(CH_2)_4C\equiv CCH_2 \diagdown \diagup (CH_2)_7CO_2H$$
$$\underset{H}{\ } \quad \underset{H}{\ }$$

2.4

2.5

2.1　Biosynthesis

The biosynthesis of fatty acids is under the control of enzymes called fatty acid synthases. Bacterial synthases typically compromise aggregates of six or seven discrete enzymes, whilst in plants and mammals, more sophisticated synthases comprise two identical, multifunctional proteins each possessing seven different catalytic activities. In addition, each synthase interacts with another protein, acyl carrier protein (ACP), which acts as a cofactor. Acyl carrier protein has a 4-phosphopantetheine moiety (cf. the structure of coenzyme A) attached to a serine residue, and the growing fatty acid chain is bound to the phosphopantetheine via its thiol group during most of the assembly process.

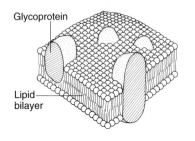

Fig. 2.1

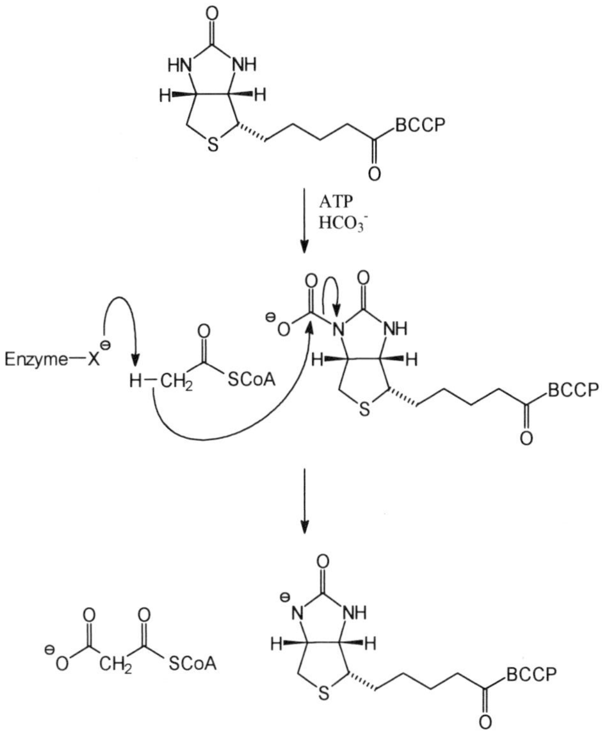

Fig. 2.2

The main building block of fatty acid biosynthesis is acetate in the form of acetyl-SCoenzyme A, though most of this is first converted into malonyl-SCoenzyme via a carboxylation reaction. The cofactor biotin and the associated biotin carboxyl carrier protein (BCCP) are involved in this process, which is depicted in Fig. 2.2. The mode of joining of an acetyl unit with a malonyl unit is conceptually very simple, and is formally an example of a Claisen ester condensation in which the carbon nucleophile is provided by a decarboxylation of malonate. The involvement of malonate and the validity of this mechanism followed the discovery that bicarbonate was essential for fatty acid biosynthesis, yet when ^{14}C-labelled bicarbonate was employed, no isotope was incorporated into the fatty acids. The actual enzyme-mediated process is somewhat more complex and is shown in Figs 2.3 and 2.4.

The two identical multifunctional proteins of the synthase each possess one site for the ACP and its pantetheine moiety and one so-called cysteine site on the condensing enzyme which is also known as β-ketoacyl synthase. In addition, there are six other sites associated with catalytic activities. These are the enzymes acetyl and malonyl transacylase, which catalyse the reaction between the cysteine thiol and acetyl-SCoA, and between malonyl-SCoA and the pantetheine thiol; beta- ketoacyl reductase; dehydratase; enoyl reductase; and thioesterase. The two proteins are arranged in a head-to-tail configuration as shown in Fig 2.3. The cycle of elongation (to form the fatty acid) begins with attachment of acetyl-SCoA to the cysteine site of one of the proteins i.e. a *trans*-thioesterification. Malonyl-SCoA is loaded on to the pantetheine site on the second protein in the same way, and the Claisen ester condensation then takes place catalysed by the β-ketoacyl synthase (condensing enzyme). The acetyl unit is thus transferred from the cysteine site on to the pantetheine site to yield a butanoyl thioester, and the rest of the chemistry occurs at that site.

The next stage involves a stereospecific reduction of the ketone carbonyl with the pro-4*S* hydrogen of NADPH acting as the source of hydride, to yield the *R*-hydroxythioester (Fig. 2.4). This then suffers dehydration which involves, somewhat surprisingly, a *syn* elimination to yield the (*E*)-unsaturated thioester. Finally, a further reduction ensues involving NADPH—the pro-*R*-hydride is transferred on this occasion—to yield the saturated

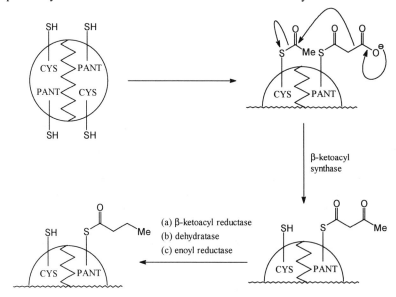

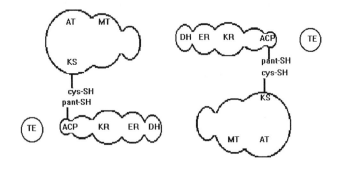

Fig. 2.3

KEY:

AT acetyltransacylase   DH dehydratase
MT malonyl transacylase   ER enoyl reductase
KS β-ketoacyl synthase   KR β-ketoacyl reductase
ACP acyl carrier protein   TE thioesterase

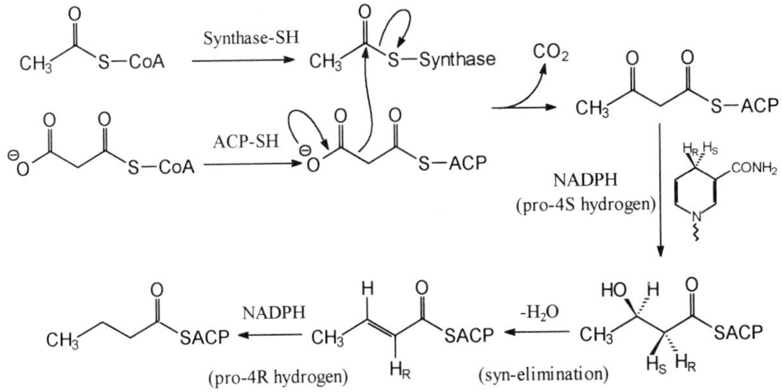

Fig. 2.4

thioester. This is then transferred to the cysteine site to await attack by a further molecule of malonyl-SCoA which initiates a further cycle of reactions.

The saturated thioester grows until it reaches around 14 to 18 carbons in length, at which point it attracts the attentions of the thioesterase and is released from the synthase. The thioesterase typically has about a 70 per cent specificity for C_{16} chains thus releasing palmitic acid, though small amounts (*c.* 20 per cent) of the corresponding C_{14} acid (myristic acid) and the C_{18} acid (stearic acid, *c.* 10 per cent) are also produced.

Two routes are known for the introduction of double bonds into these saturated fatty acids: one is used by aerobic organisms (e.g. plants and mammals), and the other is used by anaerobic bacteria. The latter route has been investigated thoroughly using stereospecifically deuterium-labelled substrates, and the mechanism of the reaction is shown in Fig. 2.5. Only monounsaturated fatty acids are produced by this route. The aerobic route is less well understood, but the mechanism shown in Fig. 2.5 is usually invoked, and polyunsaturated acids can be produced via this mechanism.

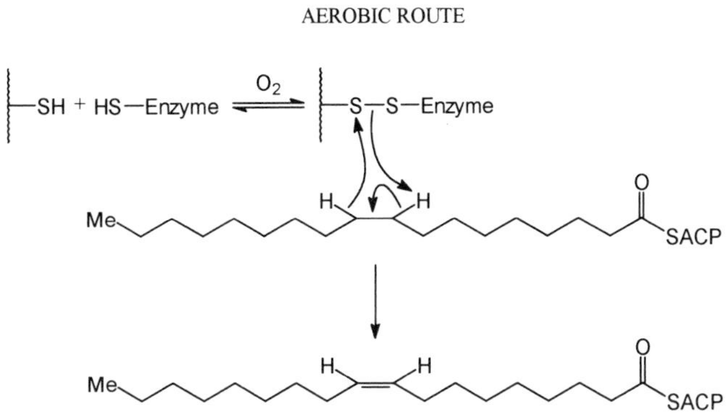

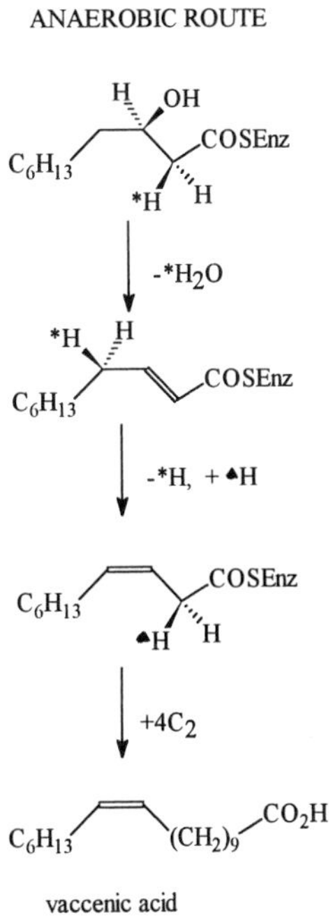

ANAEROBIC ROUTE

AEROBIC ROUTE

Fig. 2.5

2.6

2.2 Polyacetylenes

As mentioned earlier these metabolites are particularly common in members of the daisy family (Compositae and in some of the Basidiomycetes), where they probably act as anti-microbial or antiviral agents, and at least in some instances are inhibitors of seed germination. The biosynthesis of the simple polyacetylenes has been studied extensively. The routes to other more exotic naturally occurring alkynes like wyerone, **2.5** and the thiophene, **2.6** have not been established, but the heterocyclic rings are presumably formed via addition of water and H_2S (respectively) across two consecutive triple bonds.

2.3 Prostaglandins and leukotrienes

The existence of the prostaglandins was first noted in 1930 when two New York scientists found that certain unidentified substances in human semen would cause smooth muscle (e.g. the muscle of the gastrointestinal tract and the uterus) to contract. In 1934 the pharmacologists Goldblatt and von Euler confirmed this finding, and also demonstrated that the substances could lower the blood pressure of animals. Von Euler coined the term 'prostaglandins' because he believed that they originated from the prostate gland. One of his students, Bergström, finally elucidated the structures of the main prostaglandins in 1962, using milligram amounts of the compounds isolated from several tonnes of sheep vesicular glands. The synthetic chemists were then unleashed, and several hundred syntheses of the natural compounds and structural analogues have been described since 1965. Many of these analogues, and all of the principle prostaglandins, exhibit potent biological activity at the microgram level (see below for examples), but none of them has proved to be of enormous clinical utility.

The biosynthesis of the prostaglandins is shown in Fig. 2.6 and proceeds from arachidonic acid which possesses a typical 'skipped' polyolefinic structure. A free radical mechanism of fatty acid oxidation is implicated in the first step. The key enzyme, cyclooxygenase, catalyses this oxidation and the subsequent cyclization of the hydroperoxy radical to produce the endoperoxide, hydroperoxide PGG_2. This suffers several metabolic fates to produce, ultimately, the various main prostaglandins. Use of $^{18}O_2$ established that the oxygen atoms at both C-9 and C-11 of these metabolites were derived from the same molecule of oxygen.

The two metabolites PGI_2 (prostacyclin) and TXA_2 (thromboxane A_2) are of particular interest. PGI_2 was first identified in 1976 by John Vane and co-workers at the Wellcome Research Laboratories in the UK. They noted that the aggregation of blood platelets could be arrested, and even reversed, by a factor released from the walls of blood vessels. They ultimately showed that the aggregation was in fact under the control of two agents: PGI_2 (biological half-life *c.* 2 minutes) which prevented aggregation, and TXA_2 (half-life *c.* 30 seconds) which caused aggregation. TXA_2 had been isolated by Samuelsson and co-workers in 1975, but given its instability its structure was unproven until it was synthesized by Still in 1985.

As for the other prostaglandins, PGE$_2$ and PGF$_2$, they act in concert to control the contractile state of the bronchioles (small airways in the lungs); they produce contractions in the uterus at full term of pregnancy; and PGE$_2$ strongly inhibits gastric acid secretion. PGE$_2$ also encourages the formation of a protective layer of mucus on the gut wall, which helps to reduce that damaging effect of gastric acid (HCl) and the digestive enzymes. PGD$_2$, in contrast, behaves like PGI$_2$ and inhibits aggregation of blood platelets. The effect of the prostaglandins on uterine activity has led to their use, in certain African countries, as an abortifacient, and they are also used to induce labour in women who are late in going into labour.

The instability of PGI2 is obviously due to the enol ether grouping, and several stable analogues of this metabolite, e.g. Iloprost (**2.7**) have had limited success in the treatment of clinical conditions where platelet aggregates must be broken up, e.g. coronary and stroke.

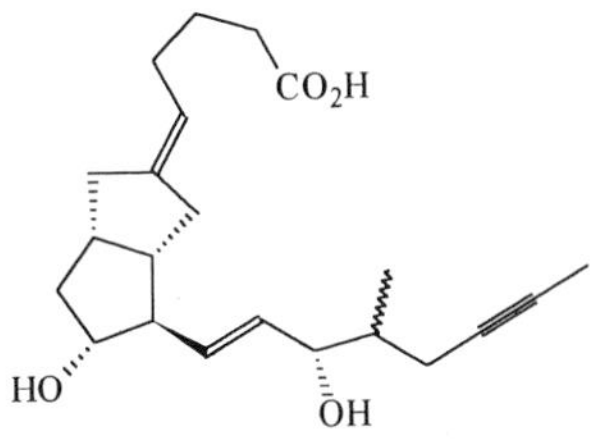

2.7

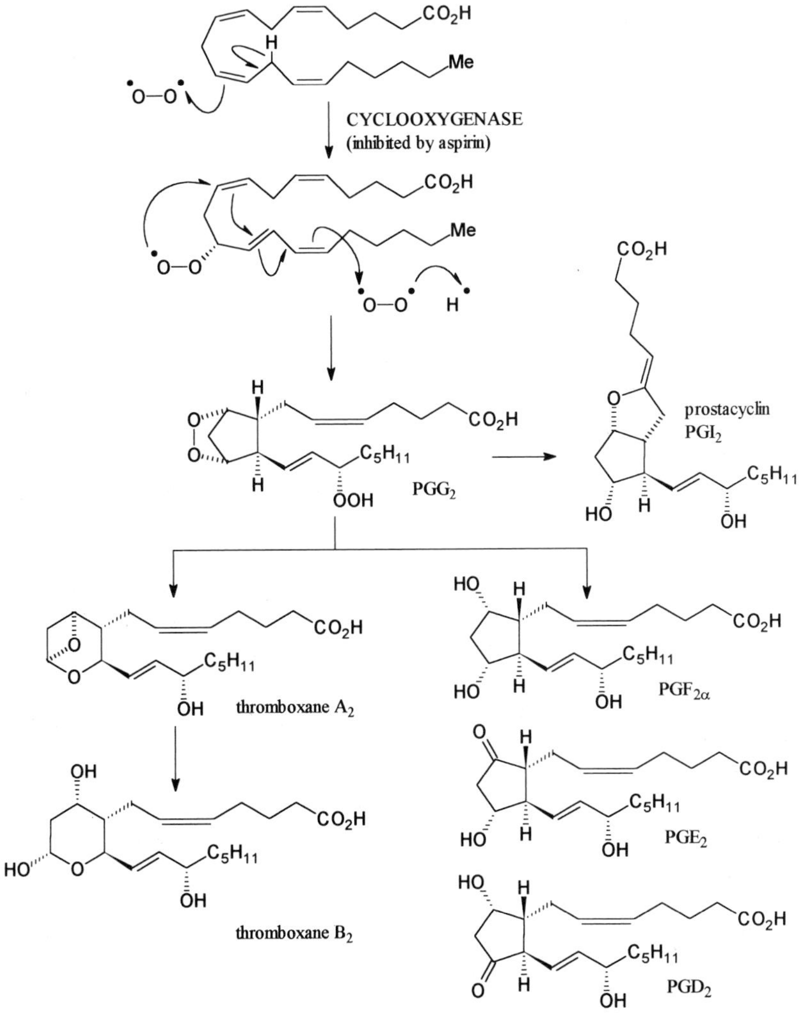

Fig. 2.6

Probably the most important prostanoid in commercial terms is the analogue fluprostenol, **2.8**, which is primarily used as a luteolytic agent for cattle—it initiates menstruation so that a farmer can synchronize the menstrual cycles of all of his cows and thus assess the exact time for artificial insemination. It has also been used to treat human infertility.

2.8

One other discovery to arise from the work of Vane and co-workers was that aspirin inhibited the key enzyme cyclooxygenase (probably through acetylation of an essential serine residue), and this immmediately provided a partial understanding of the mode of action of this widely used drug. Prostaglandins are intimately involved in the causation of pain and inflammation, hence the analgetic and anti-inflammatory potency of aspirin. It also explained why long-term usage of aspirin often causes stomach ulcers, since the cytoprotective effect of PGE_2 would be compromised.

The other major class of metabolites produced from arachidonic acid are the leukotrienes. Since the 1950s it had been known that highly inflammatory agents were produced during an allergic reaction, and until the identity of these agents was established in 1975, they were known as the 'slow reacting substances of anaphylaxis' (SRSAs). Their biosynthesis (Fig. 2.7) also commences with arachidonic acid, and this suffers autoxidation, catalysed by the enzyme 5-lipoxygenase, to produce 5-HpETE (5-hydroperoxyeicosatetraenoic acid). Loss of a proton from C-10 and movement of electrons leads to the epoxide known as leukotriene A_4, which reacts with the tripeptide glutathione to produce leukotriene C_4. Sequential loss of glutamic acid then glycine leads in turn to leukotrienes D_4 and E_4. These compounds are produced in minute amounts in response to various hormonal, chemical, or immunological stimuli, and have an array of pharmacological effects with potencies in the nanomolar range. In particular, they are implicated in various allergic diseases, especially bronchial asthma, and considerable effort has been expended in a search for inhibitors of the enzyme 5-lipoxygenase.

Arachidonic acid also gives rise to numerous other metabolites, e.g. leukotriene B_4, implicated in such diseases as psoriasis, ulcerative colitis, and arthritis; and the lipoxins, whose exact biological role is presently unclear. One might reasonably enquire why mammals produce such a plethora of apparently harmful compounds, and a partial answer can be gleaned from an examination of the role of mast cells in an allergic response. These white blood cells carry on their cell surface a special class of antibody (immunoglobulins E) which bind allergens (e.g. fungal particles, house mites, etc.). Once bound, a structural change occurs within the cell membrane causing it to become selectively permeable to calcium ions. The inrush of calcium ions then activates a number of cellular enzymes, including various lipases, and these catalyse the release of arachidonic acid from glycerides. The biosynthesis of both prostanoids and leukotrienes can now occur, and these are ultimately released from the cell, along with histamine, as a cocktail of bioactive substances. The resultant sneezing, coughing, and other more serious manifestations of an allergic response can thus be viewed as an attempt by the body to rid itself of harmful invading organisms.

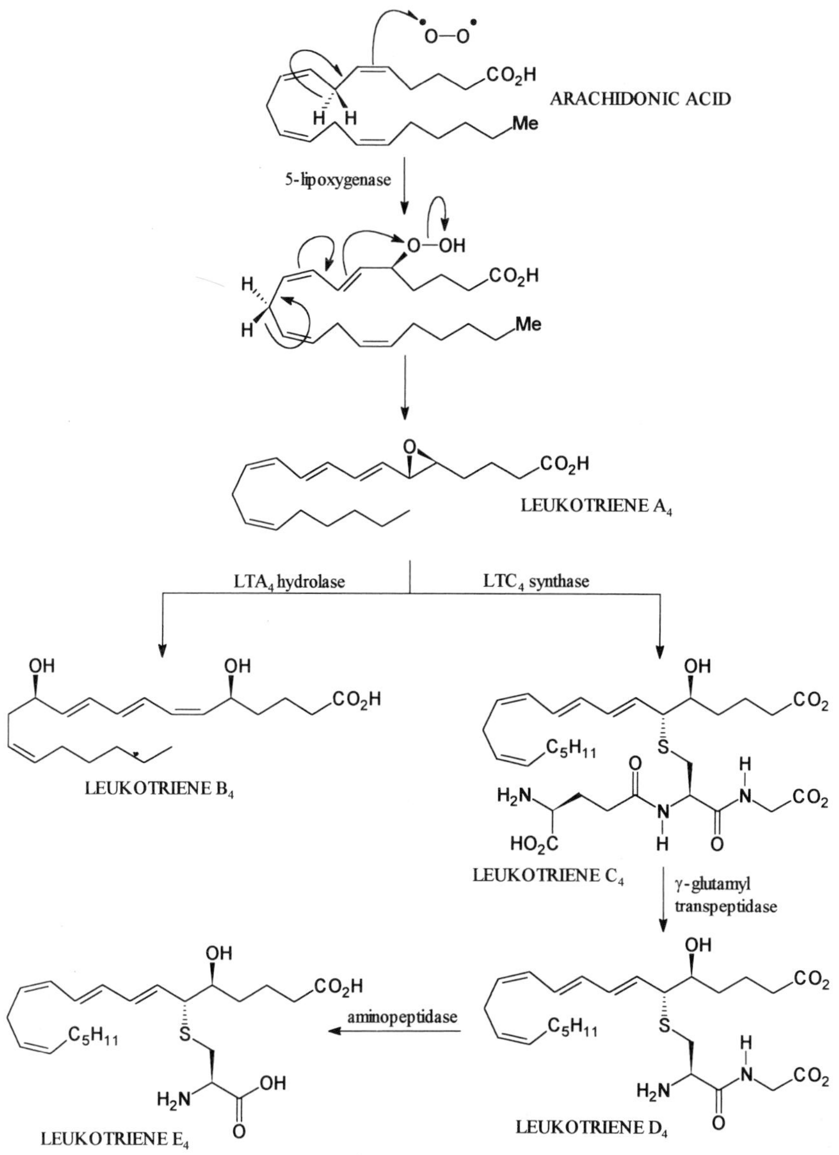

O—O
CO2H
ARACHIDONIC ACID
H
H
Me
5-lipoxygenase
O—OH
CO2H
H
H
Me
O
CO2H
LEUKOTRIENE A4
LTA4 hydrolase
LTC4 synthase
OH
OH
CO2H
LEUKOTRIENE B4
OH
CO2
C5H11
S
O
H
H2N
N
N
CO2
HO2C
H
O
LEUKOTRIENE C4
γ-glutamyl
transpeptidase
OH
CO2H
C5H11
S
H2N
OH
O
LEUKOTRIENE E4
aminopeptidase
OH
CO2
C5H11
S
H
N
CO2
H2N
O
LEUKOTRIENE D4

Fig. 2.7

2.4   Insect pheromones and other fatty acid metabolites

Most insect species employ secondary metabolites as agents of chemical communication. Where this communication is between members of the same species, these compounds are known as 'pheromones', and a few typical examples are shown in Fig. 2.8. The sex pheromones are released by females of the species when they are sexually receptive, and act as a lure to males, whilst the alarm pheromones are released by one member of a species to warn other members of the presence of a predator. Saffynol from the safflower is produced when the plant is under environmental stress or attack by a virus. Such 'stress metabolites' are known as 'phytoallexins'. These compounds (shown in Fig. 2.8) are all clearly fatty acid metabolites, and in the main they are synthesized by the insects using simpler fatty acids present in their (plant) diet. The interactions that they mediate are part of what is usually termed 'ecological chemistry', and this will be discussed in Chapter 7.

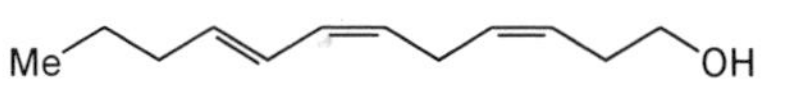

alarm pheromone of the Southern subterranean termite

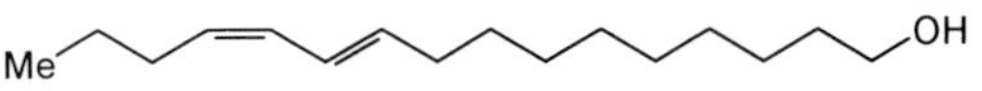

sex pheromone of the silkworm moth

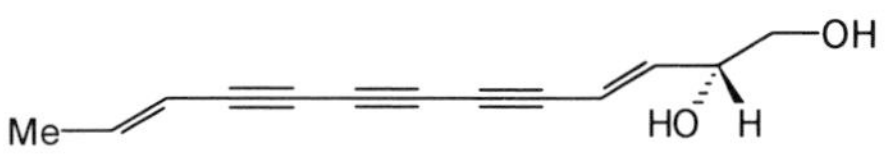

phytoallexin from the safflower

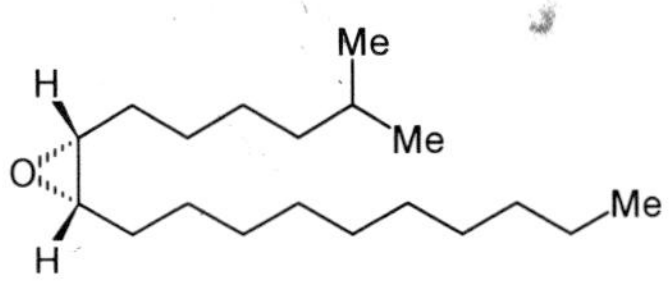

sex pheromone of the gypsy moth

Fig. 2.8

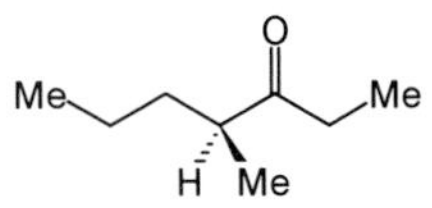

alarm pheromone of the ant *Atta texana*

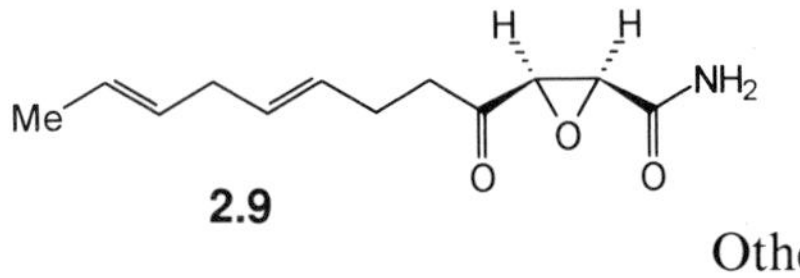

2.9

Other fatty acid metabolites that are of particular contemporary interest include cerulenin, **2.9** and brefeldin **2.10**. The former is a metabolite of the mould *Cephalosporium caerulens*, and is a potent inhibitor of fatty acid synthases. It acts specifically to block the enzyme that catalyses the reaction between the acyl thioester (at the cysteine site) and malonyl thioester (at the pantetheine site). It has been prepared in radioactive form using [14]C- and ^{3}H-acetates and is being used to probe the biochemical details of fatty acid and polyketide (see Chapter 3) biosynthesis.

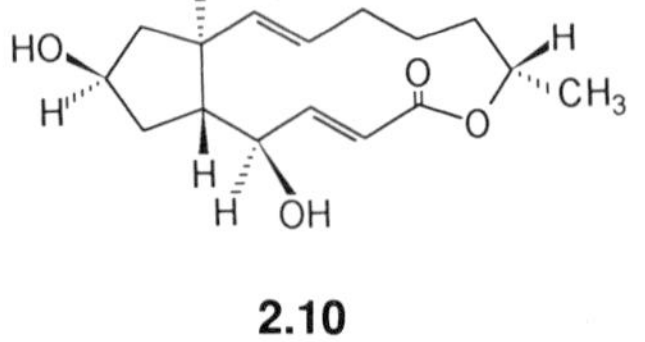

Brefeldin (**2.10**) is a metabolite of the fungus *Eupenicillium brefeldianum*, and is currently of considerable interest because it appears to inhibit the processing of antigens (eg foreign proteins of viruses) by the class of white cells known as cytotoxic T-lymphocytes. Such processing is an important prelude to the production of antibodies and an efficient immune response.

2.10

3 Polyketides

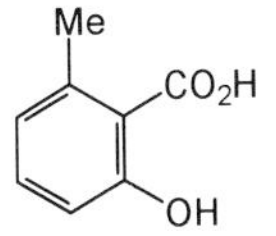

The condensation of acetyl-SCoA with malonyl-SCoA units also gives rise to numerous aromatic compounds, many with a *meta*-disposition of hydroxyl groups, but also to other cyclic compounds which do not possess aromatic rings. These are all products of the so-called 'polyketide pathway', and typical examples include 6-methylsalicylic acid, **3.1**, the antifungal compound griseofulvin, **3.2**, the mycotoxins patulin **3.3** and ochratoxin **3.4** (this also includes a moiety derived from the amino acid phenylalanine), and the macrocyclic antibiotic erythromycin A, **3.5** (see Fig. 3.14). Most of these are products of moulds and other soil and airborne microorganisms, and are especially abundant in bacteria of the family actinomycetes.

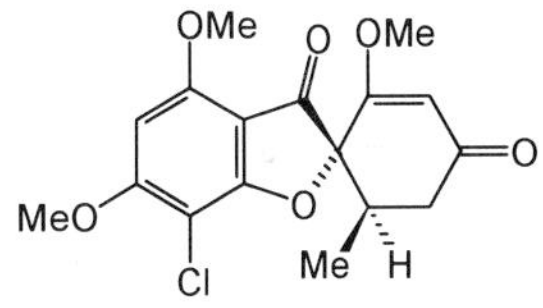

An early insight into how these compounds might be assembled was provided by Collie in 1907, when he treated heptane-2, 4, 6-trione with base to produce a naphthalene derivative via a series of intermolecular aldol reactions (eqn 3.1). This biomimetic synthesis using a triketide was the spur

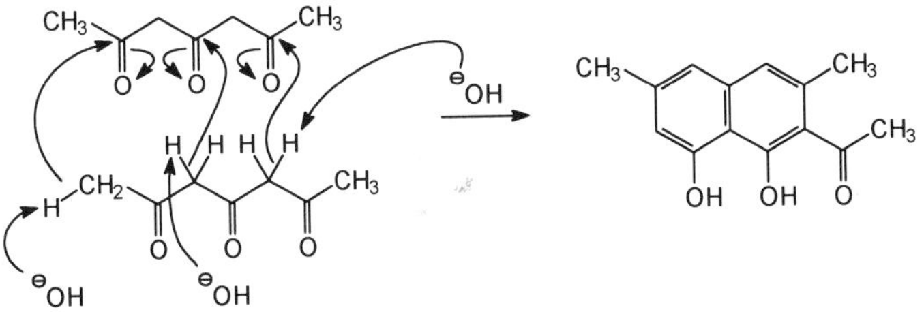

$$(3.1)$$

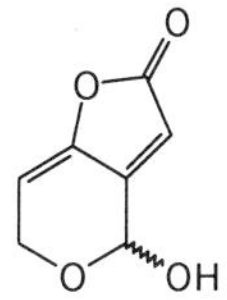

for others to formulate a 'polyketide hypothesis' that envisaged the biosynthesis of polyphenols from polyketomethylene intermediates. It was the pioneering work of Arthur J. Birch in the 1950s that established the reality of this hypothesis. He used the then newly available 1- and 2-^{14}C-labelled acetates for feeding experiments with the mould *Penicillium patulum*, and through degradation experiments, showed that 6-methylsalicylic acid was derived from four moles of acetate. It was only later realized that one acetate and three malonates were involved (Fig. 3.1).

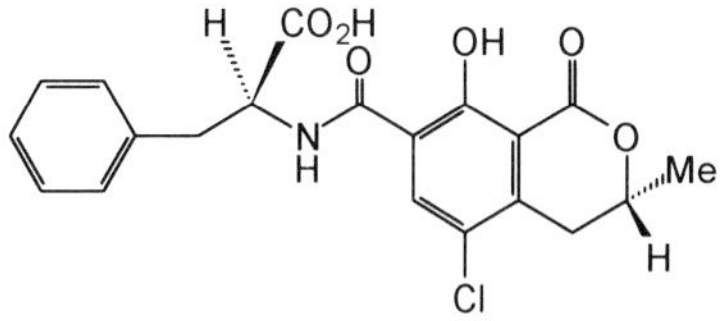

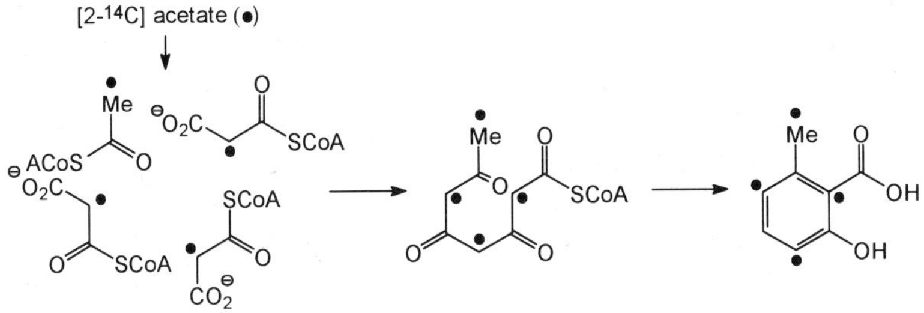

Fig. 3.1

3.1

3.2

3.3

3.4

The assembly of the metabolites is under the control of enzymes known as polyketide synthases. These have been classified into three types. Types I and II, found in bacteria and fungi, possess multifunctional enzymes or aggregates of monofunctional enzymes that act upon substrates that are bound by thioester linkages to an acyl carrier protein. Type III synthases, found in plants, lack the ACP moiety and use coenzyme A esters. Much work has been carried out over the past five years to isolate and sequence the genes for both fatty acid synthases and the polyketide synthases. It is now clear that some of the PK synthase genes have a strong sequence similarity to the FA synthase genes, reflecting the similar roles for the enzymes coded for by the genes. Other genes like those coding for the polyketide cyclases, which catalyse formation of the aromatic rings, have, not surprisingly, no counterparts amongst the genes coding for the fatty acid synthases.

The pathway envisaged addition of successive two-carbon units, generally without the prior reduction of the existing carbonyl group, as is observed in the fatty acid pathway. Other work has substantiated this proposal, and in fact the plethora of structures which are of polyketide origin owe their diversity to the fact that the carbonyl groups may or may not be reduced during biosynthesis. In addition, when reduction occurs, the resultant alcohols may or may not be eliminated, so the final metabolites may contain a complex mixture of functional groups including keto, hydroxyl, or alkene wherever a carbonyl group existed in the polyketide progenitor. As we shall see later, the possible involvement of propionate, butanoate, and other starter units in place of acetate adds further complexity.

3.1   The use of ^{13}C-NMR and related techniques

Obviously the degradative work carried out by Birch and others was tedious and relatively inefficient, and the introduction of the ^{13}C-isotope in the 1970s revolutionized both structure determination and biogenetic investigation. This carbon isotope has a natural abundance of 1.1 per cent, and since it possesses a nuclear spin $I = 1/2$, NMR signals may be observed provided a sensitive spectrometer is available. With the advent of pulsed Fourier-transform spectrometers, the acquisition of ^{13}C-spectral data has become routine, and a natural abundance spectrum of most compounds can be obtained given a few milligrams of material.

One great advantage of ^{13}C-NMR is the considerable spread of chemical shifts. Compare, for example, the ^{1}H and ^{13}C spectra for the steroid androst-4-en-3, 17-dione (Spectra 3.1 and 3.2). The former is relatively uninformative, but almost all of the carbon signals can be clearly identified in the ^{13}C

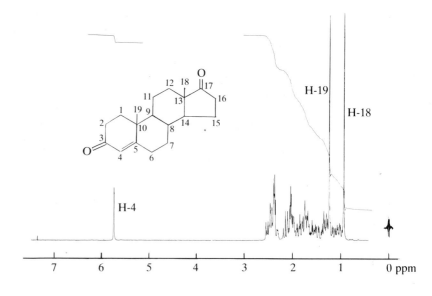

Spectrum 3.1 ^{1}H-NMR at 220 MHz in CDCl$_3$

Peak	1	2	3	4	5	6	7	8	9	10	11	12
ppm	220.07	198.96	170.25	123.98	78.72	77.29	75.86	53.71	50.72	47.38	38.56	35.64
Assignment	C – 17	C – 3	C – 5	C – 4		CDCl₃		C – 9	C – 14	C – 13	C – 10	C – 16 & C – 1

Peak	13	14	15	16	17	18	19	20	21	22
ppm	35.05	33.85	32.48	31.21	30.66	21.66	20.26	17.30	13.66	0
Assignment	C – 8	C – 2	C – 6	C – 7	C – 12	C – 15	C – 11	C – 19	C – 18	

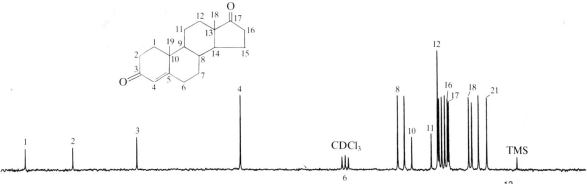

Spectrum 3.2 ¹³C-NMR (proton decoupled) at 22.5 MHz in CDCl₃

spectrum. Spectrum 3.2 has been proton-decoupled, which means that the coupling between the ^{13}C and ^{1}H atoms is not observed. Actual observation of these couplings provides further structural information, since (as expected) a methyl carbon appears as a quartet, a methylene carbon as a triplet, and a methine carbon as a doublet.

Feeding experiments using ^{13}C-labelled acetate (or malonate, or propionate, etc.) with up to 99.9 per cent enrichment in ^{13}C can lead to isotopically enriched metabolites that give NMR spectra in which certain of the resonances are enhanced while others are not. If the natural abundance spectrum has been assigned, the distribution of tracer may be discerned immediately without recourse to degradative chemistry.

Another technique that has been used to great effect is the administration of doubly labelled acetate, i.e. [1, 2–^{13}C]-acetate. In this way all carbon atoms of the metabolite are labelled in feeding experiments, but since adjacent ^{13}C nuclei will exhibit ^{13}C–^{13}C coupling, some resonances will be split into doublets in the NMR spectrum. By comparing the observed coupling constants, the adjacent nuclei can be identified, and can be assumed to derive from an intact acetate unit via malonate.

Thus, a chain of carbon atoms derived from [1, 2–^{13}C]-acetate will exhibit the labelling pattern

$$\text{Me}-\text{CO}_2^- \longrightarrow \quad -\overset{}{\underset{1}{C}}-\overset{}{\underset{2}{C}}-\overset{}{\underset{3}{C}}-\overset{}{\underset{4}{C}}- \quad \text{and} \quad -\overset{}{\underset{1}{C}}-\overset{}{\underset{2}{C}}-\overset{}{\underset{3}{C}}-\overset{}{\underset{4}{C}}-$$

The signals due to C-1 and C-2 will appear as pairs of doublets with identical J values ($J_{1,2}$). Similarly, C-3 and C-4 will provide doublets with the same value, $J_{3,4}$, and $J_{1,2}$ will not usually equal $J_{3,4}$. However, since it is statistically unlikely that labelled acetate units will be incorporated consecutively during biosynthesis (there will usually be many unlabelled acetate molecules in an organism, and to ensure this, sometimes unlabelled acetate is fed in conjunction with labelled acetate), there should be negligible coupling between C-2 and C-3. Sets of coupled carbon atoms are thus identifiable, and such results provide evidence for intact incorporation of acetate units, and for the way in which the polyketide chains are folded.

A yet more subtle use of isotopic labelling coupled with NMR investigation involves the use of acetate labelled with deuterium (^{2}H), nitrogen-15 (^{15}N), or oxygen-18 (^{18}O). Deuterium has a nuclear spin $I = 1$ and a natural abundance of 0.016 per cent, and produces broad signals with resonance frequencies and coupling constants of only about one-sixth those of hydrogen. In consequence, its main use has been in conjunction with ^{13}C. In the ^{13}C spectrum, an upfield shift of the ^{13}C signal is observed if the deuterium is directly attached (α effect), and a small upfield shift if the deuterium is one carbon atom removed from the ^{13}C atom (the β effect). Shifts are also seen when the ^{15}N or ^{18}O isotopes are attached to a ^{13}C atom.

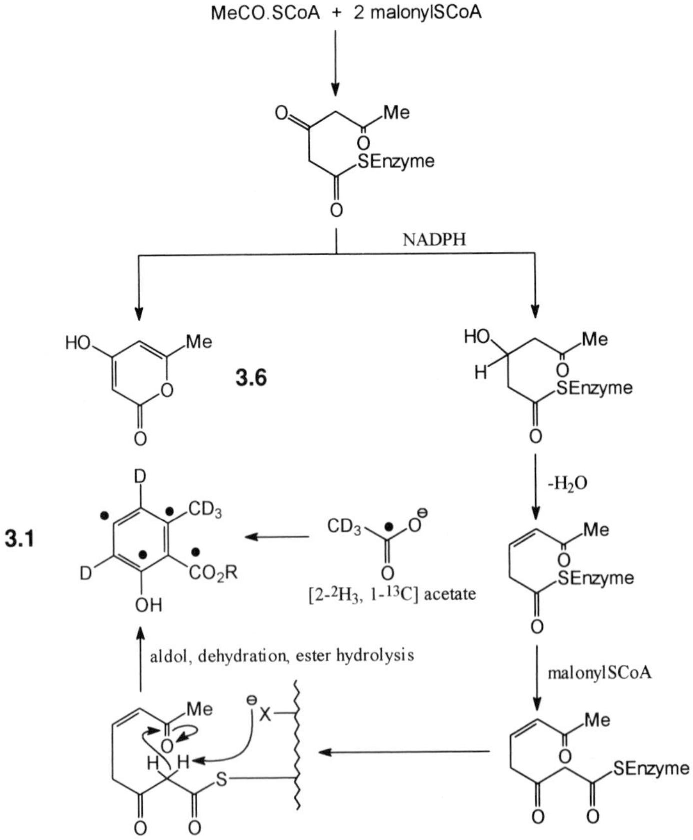

Fig. 3.2

If we return to the more recent studies of 6-methylsalicylic acid biosynthesis, the use of these techniques can be illustrated. The molecule is assembled from one molecule of acetyl-SCoenzyme A and three molecules of malonyl-SCoenzyme A, and the pathway shown in Fig. 3.2. is in accord with the results obtained from feeding experiments with variously labelled acetates. The enzymes catalysing the various steps have been isolated (and at least partially characterised) from *Penicillium patulum*.

The information available from one labelling experiment (carried out by Staunton and Abell) is depicted in Fig. 3.2 and also in Spectrum 3.3. When $[2\text{-}^2H_3, 1\text{-}^{13}C]$-acetate was employed, four enhanced carbon signals were observed in the ^{13}C-NMR spectrum, but each of these was also accompanied by satellite resonances indicating the presence of one or more deuterium atoms attached to the adjacent carbon atom (the β effect). The levels of deuterium in the various molecules of the metabolite in the NMR tube reflect the extent of deuterium loss via enolization (see Fig. 3.3) during the biosynthesis. It is interesting to note that integration of the various signals indicated that typically less deuterium was lost from C-3 and C-7 than from the other centres. This suggests that reduction and dehydration occur early in the biogenesis, thus ensuring that the CD_2 group destined to become C-3 is adjacent to a ketone group for a relatively short period. If the enzyme is starved of NADPH, the lactone **3.6** (see Fig. 3.2) is the major product.

Even more subtle experiments employing stereospecifically labelled malonates (i.e. 2*R* or 2*S*-deutero) have been used recently to demonstrate that the loss of deuterium (or hydrogen under natural conditions) is actually stereospecific, presumably reflecting the precise three-dimensional arrangement of substrate, cofactor, and interactive amino acid residues at the active sites of the various enzymes.

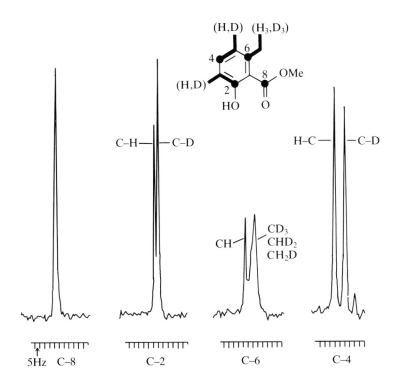

Spectrum 3.3

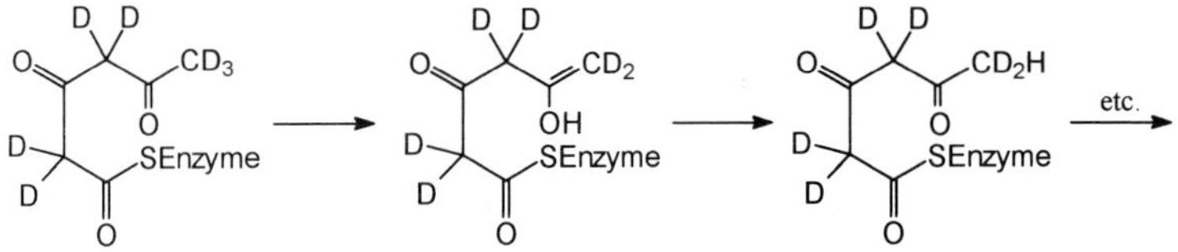

Fig. 3.3

The use of [1, 2-^{13}C]-acetate is exemplified by experiments carried out variously with *Penicillium griseum* and *P. baarnense* on the biosynthesis of penicillic acid, **3.7** (see Fig. 3.4). This interesting metabolite has antibacterial activity but is also carcinogenic. It provides a good example of the general phenomenon that secondary metabolites are often the result of a regular biosynthetic pathway followed by a (gross) structural modification. In this instance the pathway proceeds via the common mould metabolite orsellinic acid, **3.8** (see Fig. 3.4). In Fig. 3.4 the bold lines indicate C–C bonds that originate from intact ^{13}C–^{13}C units of [1, 2-^{13}C]-acetate or malonate, as established by the ^{13}C–^{13}C couplings observed in the ^{13}C-NMR spectrum of isolated orsellinic acid. Thereafter the pathway involves attack on two occasions by oxygen, probably in the form of peroxide, with subsequent decarboxylation and a pseudo-Baeyer–Villiger rearrangement to produce penicillic acid. The evidence for this mechanism is provided by the two residual ^{13}C–^{13}C couplings observed for C-2 to C-3 and C-5 to C-7, indicating that these two C–C bonds remain intact throughout the rearrangement sequence. The extra carbon atom present in penicillic acid (methyl ether carbon) is derived from the cofactor S-adenosylmethionine (SAM) (see Fig. 1.5).

It is interesting to speculate that pathways to metabolites like 6-methylsalicylic acid and orsellinic acid, which are anaerobic (i.e. the pathways do not require oxygen), are old in evolutionary terms. When the atmosphere of the Earth became more oxidizing, the moulds responded with additional pathways that used oxygen.

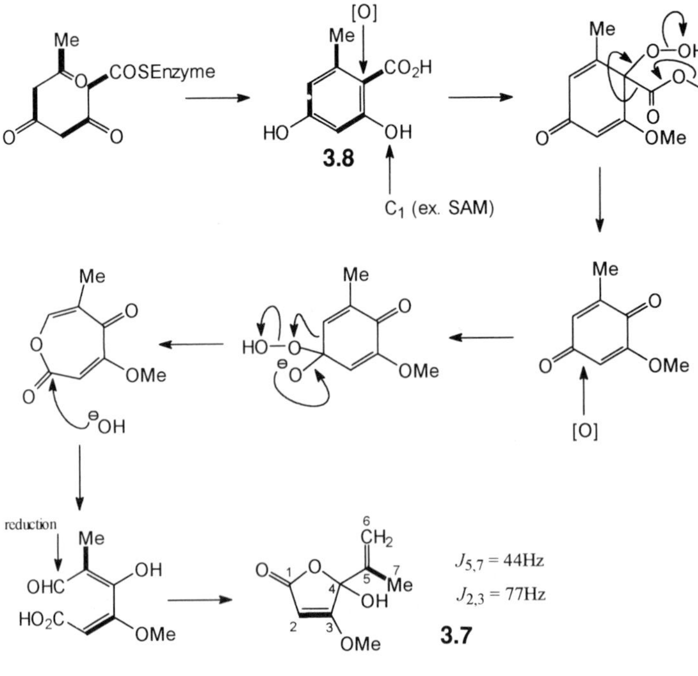

Fig. 3.4

3.2   Biosynthesis of more complex polyketide metabolites

All of the other metabolites to be described in this chapter are biosynthesized via putative polyketide intermediates, and it is convenient to discuss them according to the number of C_2 units from which they are derived. Hence citrinin, **3.9**, probably arises from the pentaketide, **3.10** via the pathway shown in Fig. 3.5. This metabolite is one of a large number of naturally occurring mould metabolites that are known to be a hazard to health. It was first isolated in 1931 from *Penicillium citrinum*, but is also a constituent of other species of *Penicillium* and *Aspergillus* moulds. These moulds grow well on rice and can give rise to the so-called 'yellow rice disease'. Toxicological tests in rats and mice have demonstrated that citrinin is markedly nephrotoxic (toxic to the kidneys), but the health hazard for humans is unknown.

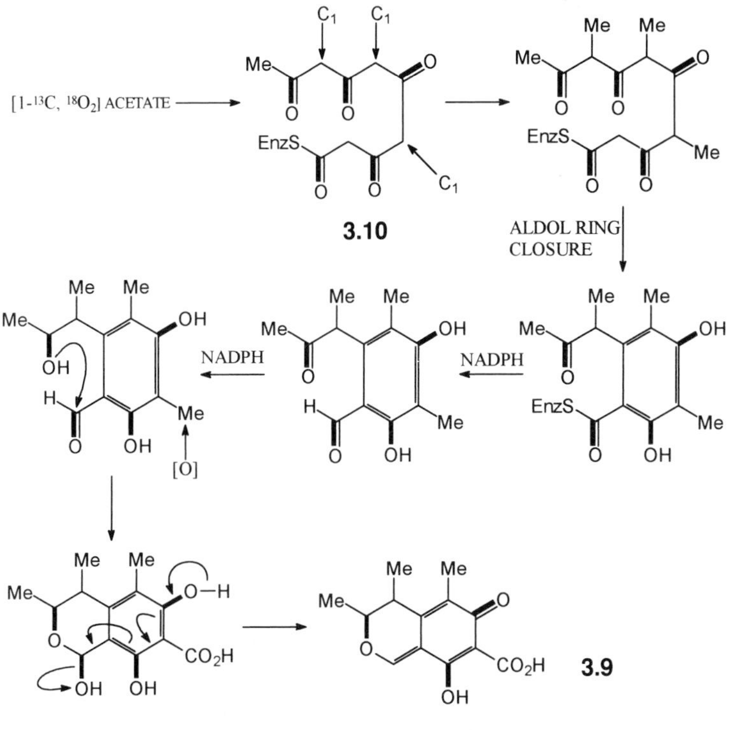

Fig. 3.5

Various labelling experiments have been carried out with *P. citrinum*, and the one shown in Fig. 3.5 involved use of [1-^{13}C, ^{18}O$_2$]-acetate. The bold lines indicate intact ^{13}C–^{18}O bonds that were established by the presence of satellite ^{13}C signals in the NMR spectrum. A number of extra methyl groups are added during the biogenesis, and again these are derived from the cofactor S-adenosylmethionine.

The lactones mellein, **3.11**, and aspyrone, **3.12** (Fig. 3.6), are but two representatives of the rich variety of metabolites produced by the mould *Aspergillus melleus*. The biogenesis of mellein is straightforward, and the data from a feeding experiment using [1,2-^{13}C]-acetate are shown in Fig. 3.6.

LIVERPOOL
JOHN MOORES UNIVERSITY
AVRIL ROBARTS LRC
TITHEBARN STREET
LIVERPOOL L2 2ER
TEL. 0151 231 4022

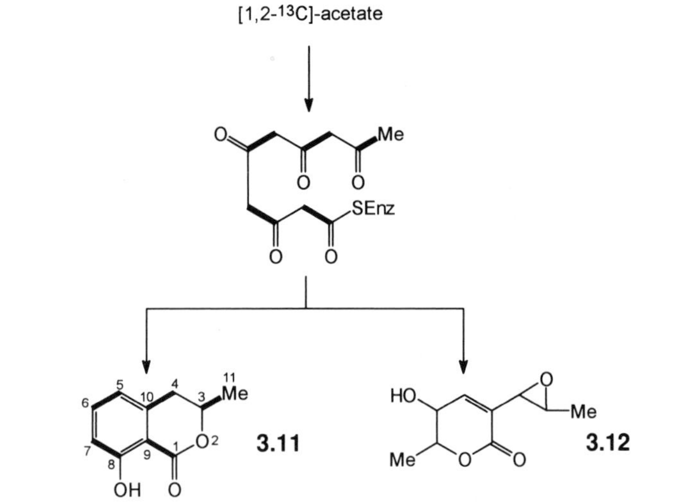

Fig. 3.6

carbon	1	3	4	5	6	7	8	9	10	11
δ_C	169.75	76.0	34.6	117.8	135.9	116.1	162.1	108.2	139.2	20.7
J/Hz	68	40	41	55	55	67	67	68	41	40
multiplicity	s	d	t	d	d	d	s	s	s	q

The pathway to aspyrone, like that to penicillic acid, involves bond cleavage and rearrangement, and the interested reader is referred to the original literature (cited at the end of book) for a discussion of the process.

The non-aromatic metabolite variotin, **3.13** (see Fig. 3.7), is produced by the mould *Paecilomyces varioti*. It possesses quite marked antifungal activity and this property allows the species to compete effectively with other microorganisms. A labelling study using [1-^{13}C]-acetate provided the labelling pattern shown in Fig. 3.7 and suggested the involvement of a hexaketide intermediate. The pyrrolidinone ring is probably derived from the amino acid proline.

Variotin has never been used as an antifungal agent in humans, but the mould metabolite griseofulvin, **3.2**, is widely used for treatment of ringworm infections in both animals and humans. During the 1930s a conifer planting programme on Wareham Heath in Dorset had to be abandoned because the saplings failed to thrive. Subsequent investigation of soil samples by the ICI Plant Protection Group at Jealotts Hill in Berkshire established that the toxicity was due to the presence of large numbers of *Penicillium* species. A predominant species was *Penicillium janczewskii* and this produced griseofulvin, a compound that had been previously isolated from a different mould, *Penicillium griseofulvum* by a group at the London School of Hygiene and Tropical Medicine. The chemical structure was not established until 1952 (by Grove and co-workers), and although it proved to be too expensive for use as an agricultural fungicide, it proved to have excellent oral fungicidal activity in humans and domesticated animals. It was introduced for this purpose in 1959 by ICI (as the drug Fulcin) and by Glaxo (Grisovin).

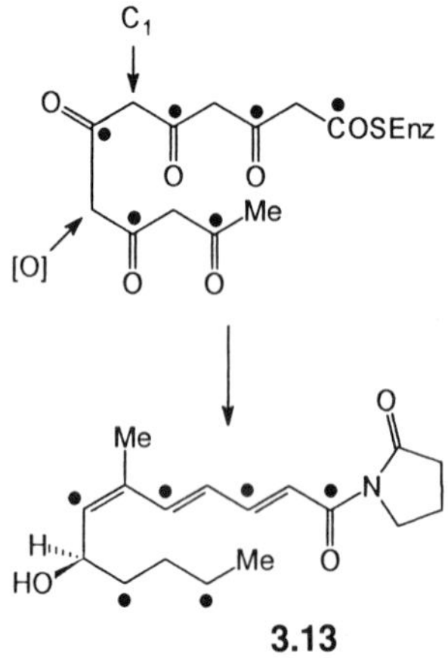

Fig. 3.7

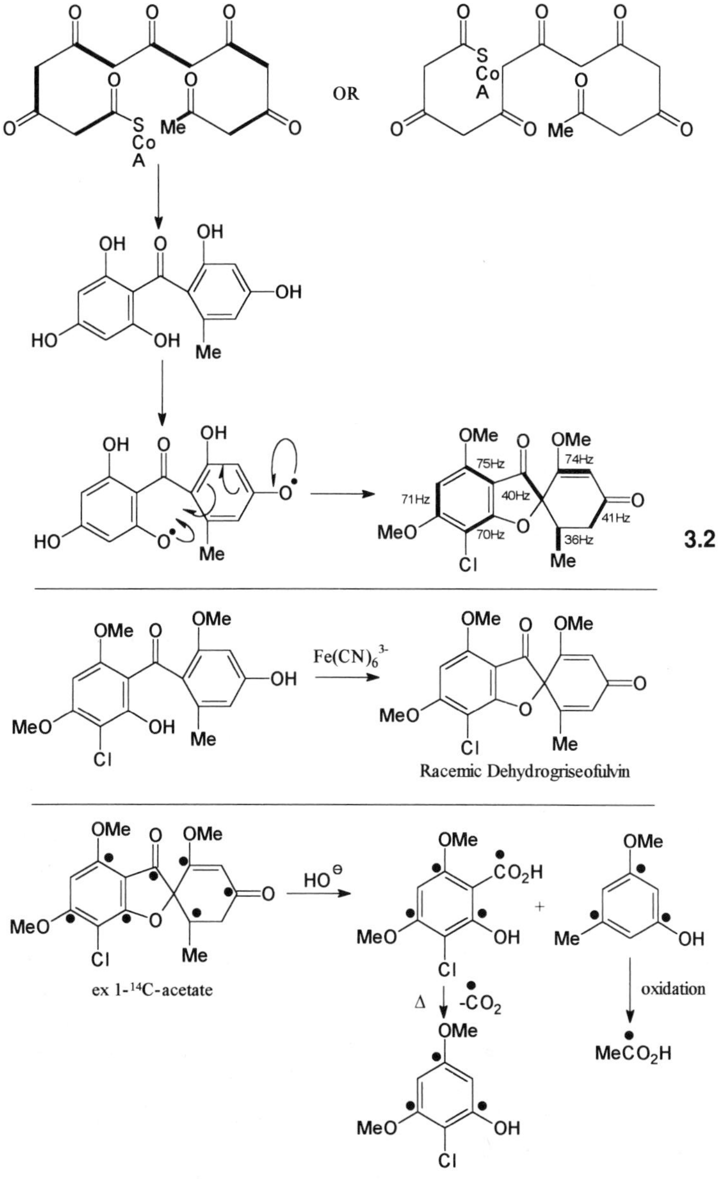

Fig. 3.8

Its biosynthesis was studied by Birch in the 1950s using ^{14}C-labelled substrates, and part of his degradation scheme is shown in Fig. 3.8. His results were confirmed and extended by various research groups in the 1970s. The ^{13}C-NMR data are consistent with the pathway shown in Fig. 3.8, and it is interesting to note that two modes of folding are possible for the putative heptaketide intermediate. The existence of coupling between all of

3.14, R = H

Fig. 3.9

the possible sets of C_2 units in the A ring suggests that both modes of folding are utilized (only one set is shown in the figure). The viability of the postulated oxidative phenolic coupling has been demonstrated using the model (biomimetic) transformation shown in Fig. 3.8.

The fungal metabolite alternariol (**3.14**, R = H; see Fig. 3.9) is particularly interesting because it has been both the subject of a thorough biosynthetic investigation and synthesized by a biomimetic route that parallels the proposed biosynthetic route. Simpson has provided definitive evidence for the route shown in Fig. 3.9 through the use of [1-^{13}C, ^{18}O$_2$]-acetate. In contrast, Staunton and Abell synthesized the benzyl-γ-pyrone **3.15** (see Fig. 3.10), which upon treatment with sodium hydroxide in aqueous methanol provided alternariol methyl ether (**3.14**, R = Me) in 80 per cent yield after acidic work-up (Fig. 3.10). The reaction presumably proceeds via the intermediate **3.16** (Fig. 3.10), which is also a likely intermediate on the biosynthetic pathway.

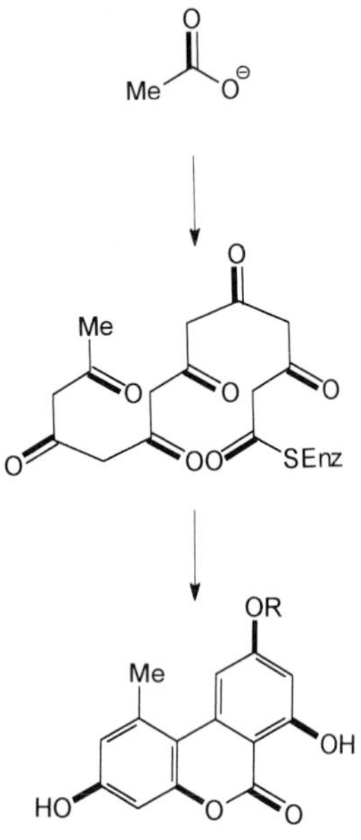

3.15

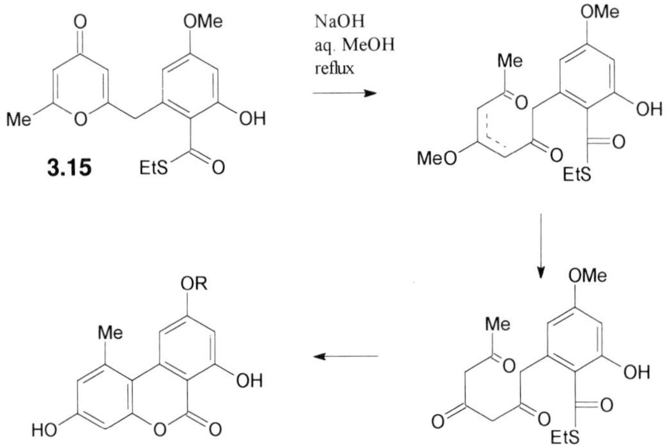

3.14 R =Me   **Fig. 3.10**   **3.16**

As a final example of a metabolite derived from a heptaketide intermediate, the unusual *N*-oxide coccinelline, **3.17** (see Fig. 3.11), is of considerable interest. It is one of a number of similar malodorous compounds released by the ladybird as part of its defensive secretion. Although not obviously of polyketide origin, the appropriate ^{13}C-labelling experiments have been carried out, and the pathway shown in Fig. 3.11 is entirely reasonable. The source of the nitrogen atom is not known. The characteristic colouration of the ladybird is a warning signal that it can defend itself by chemical means, and there are numerous other examples of insects and reptiles that advertise the noxious nature of their secondary metabolites. This kind of 'advertisement' is known as aposematism.

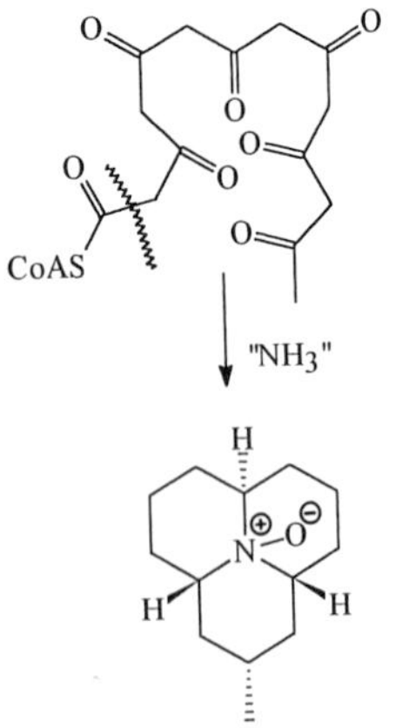

One metabolite derived from an octaketide is of particular significance, and this is the antibiotic actinorhodin, **3.18** (see Fig. 3.12), from *Streptomyces coelicolor*. This dimeric isochromanequinone has been the focus of considerable research at the molecular biological level. The cluster of genes

3.17

Fig. 3.11

3.19　　　　　　　　**3.18**　　　　**Fig. 3.12**

that supply the blueprint for the production of the enzymes involved in its biosynthesis has been isolated in the form of a continuous stretch of DNA. It has been possible to identify which parts of the DNA are required for the production of enzymes catalysing the various condensation, reduction, and dehydration reactions. Several of the genes have been sequenced, and so it is possible to predict the linear amino acid sequences of the corresponding enzymes. On the macromolecular level, the co-occurring antibiotic kalafungin, **3.19** (see Fig. 3.12), has been synthesized in ^{14}C-labelled form, and the bacterium will incorporate this compound with great efficiency (*ca.* 17 per cent incorporation of radiolabel) into actinorhodin. This suggests that the overall pathway shown in Fig. 3.12 may operate.

Intense research activity has also been associated with other polycyclic aromatic compounds, and the most important of these are the aflatoxins, e.g. aflatoxin B$_1$ (**3.20**, R = H). These were first isolated in 1960 following the deaths of thousands of turkeys which had been fed mouldy peanut meal. Various *Aspergillus* moulds were present, most notably *Aspergillus parasiticus* and *A. flavus*, and the consumed aflatoxins caused gross liver damage to the turkeys. In fact there is now abundant evidence that in those parts of the world where peanut meal is a staple part of the diet, human populations suffer from elevated levels of cirrhosis and liver cancer, presumably due to the consumption of the aflatoxins over a period of many years. In addition, cows that consume gross or fodder contaminated with aflatoxin B$_1$ convert this into aflatoxin M$_1$ (**3.20**, R = OH), and this is believed to be the causative factor of childhood cirrhosis in India. Many other mycotoxins represent real health hazards for human populations, and other examples will be discussed in due course. The biosynthesis of the aflatoxins from a decaketide intermediate is highly complex, but further information can be found in the references cited at the end of the book.

Finally, brief mention should be made of the numerous metabolites that are biosynthesized from a starter unit other than acetyl-SCoenzyme A. Many of these are derived from the simple homologue propionyl-SCoA, **3.21** (see Fig. 3.13) and bacterial metabolite daunomycin, **3.22**, is produced in this way (Fig. 3.13). This compound and several semi-synthetic analogues are broad-spectrum antitumour agents, and are used with great effect for the treatment of, in particular, leukaemia, lymphomas, and many other tumours.

3.20

LIVERPOOL JOHN MOORES UNIVERSITY
LEARNING SERVICES

The non-aromatic metabolite erythromycin A, **3.5**, is also derived from propionate units (Fig. 3.14). The successful incorporation of the thioester mimic, **3.23** in ^{13}C-labelled form supports previous suggestions that reductions (and probably dehydrations) occur concomitantly with chain elongation, rather as in fatty acid biosynthesis. The ^{13}C-NMR spectrum showed the expected satellite peaks observed when [1-^{13}C, 1-^{18}O$_2$]-acetate was employed in a feeding experiment. Erythromycin A is a structurally simple member of a class of metabolites known as macrolide antibiotics. It is produced by the bacterium *Streptomyces erythreus*, and is now widely prescribed for the treatment of penicillin-resistant bacterial infections.

Analysis of the polyketide synthase genes of *Saccharopolyspora erythrea*, which also produces erythromycin and its precursor 6-deoxyerythononlide-B (**3.24**), is now far advanced. Three large genes that code for three polyfunctional proteins have been identified and sequenced. As a result it is clear that the growing polyketide chain is processed by a series of enzymes (ketosynthases, acyltransferases, ketoreductases, dehydrogenases, ACPs and one thioesterase) to produce initially (**3.24**) and thence erythromycin A. The possible implication of this research are both exciting and far-reaching, since rational alteration of the gene sequences could provide modified synthase activity with resultant production of novel antibiotics. Extension of this type of study to other organisms producing polyketide metabolites will be a major area of endeavour in the next few years. A whole range of novel, biologically active (and potentially clinically useful) compounds will surely be identified.

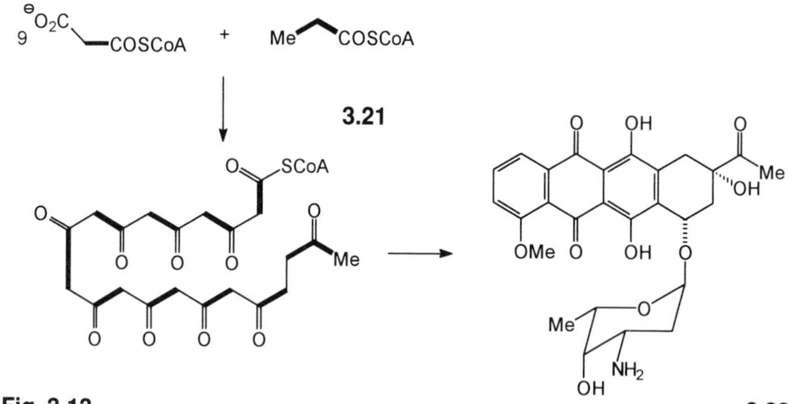

Fig. 3.13

These compounds provide good examples of the complexity that must reside in enzyme systems involved in biosynthesis from polyketide intermediates. In particular, there must be precise control of the level of oxidation at different points throughout the biosynthesis, and the wealth of structural types belies the apparent simplicity of the organisms producing them.

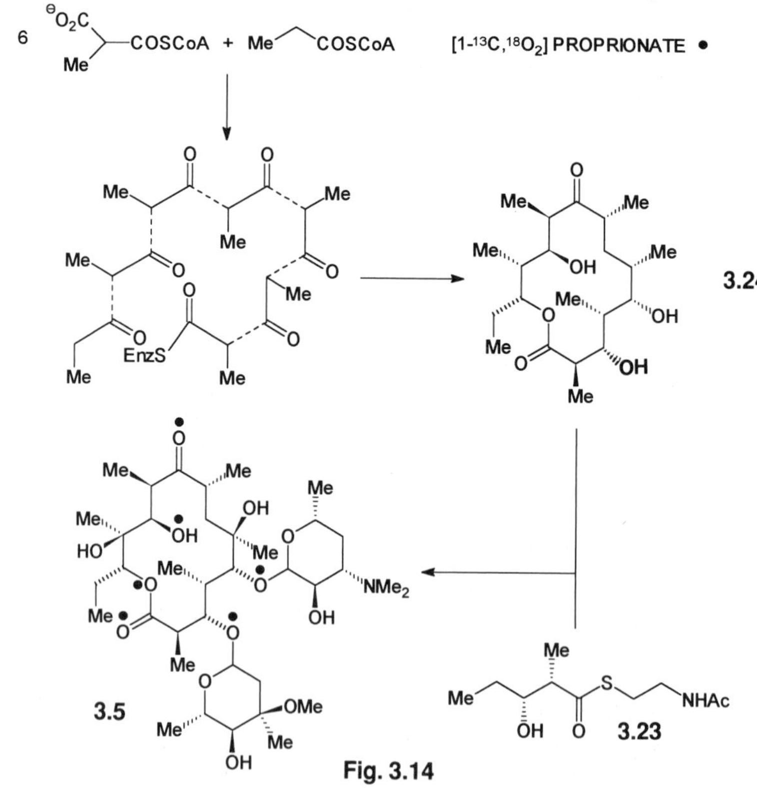

Fig. 3.14

4 Terpenoids and steroids—the isoprenoids

Acetyl-SCoenzyme A is also the main building block of the terpenoids and steroids. This large family of natural products includes such diverse structures as α-pinene (from pine oil) **4.1**; thujone, a major (psychoactive?) constituent of the banned liqueur absinthe, **4.2**; artemisinin, a potent antimalarial compound from Chinese *Artemisia annua*, **4.3**; cholesterol, **4.4**; and β-carotene, **4.5**, the major accessory pigment (to chlorophyll) in photosynthesis. When the structures of the simple ten-carbon compounds (monoterpenes) were established in the nineteenth century, it quickly became apparent that they contained an integral number of five-carbon units, as depicted for **4.1** and **4.2**. Indeed, when monoterpenes were pyrolysed they often gave rise to isoprene (2-methylbutadiene) as a major decomposition product, and biogenetic speculation involving isoprene as a building block was often useful in structural studies:

This led to the formulation of the 'isoprene rule' (hence the collective name 'isoprenoids'), and later to Ruzicka's 'biogenetic isoprene rule' (1953). He recognized that isoprenoids were derived from an integral number of biological equivalents of isoprene, joined together in a head-to-tail fashion, with subsequent modification of structure due to rearrangements and functional group changes. The identity of this biological five-carbon unit was unknown until the mid-1950s, when it was discovered that a component of so-called 'brewers' solubles' (the liquor left after the beer has been decanted) could support the growth of a mutant strain of bacteria that required acetate for growth. This bacterium thus produced various isoprenoids only if it was supplied with acetate or this new compound, christened mevalonic acid, **4.6** (a six-carbon compound) (see Fig. 4.1). In fact it was subsequently shown that only the 3 (*R*)-form of mevalonic acid (MVA) could act as a precursor of isoprenoids, and that if administered in ^{14}C-labelled form to yeast cells, mammalian liver extracts, etc., it was converted into cholesterol with loss of CO_2.

4.1   Biosynthesis of mevalonic acid

The elucidation of the biosynthetic pathway to cholesterol was one of the triumphs of the 1960s, and was largely due to the efforts of Bloch, Lynen, Cornforth, and Popjak, who shared the Nobel Prize for their discoveries in 1964. Incorporation studies were carried out with 1- and 2-^{14}C-acetates, 2-^{14}C-MVA, and then with stereospecifically labelled tritio-MVA species, and involved extensive degradative experiments with the radiolabelled samples of cholesterol and various isolated intermediates. As a result of these investigations and more recent enzyme studies, the pathway to the key five-carbon units can be described in detail (Fig. 4.1).

The initial step, catalysed by the enzyme acetoacetyl-SCoA thiolase, involves a Claisen ester condensation between two molecules of acetyl-SCoA. The second step, catalysed by the enzyme hydroxymethylglutaryl-SCoA (HMG-SCoA) synthase, is formally an aldol reaction, and the subsequent reduction of HMG-SCoA, **4.7** (See Fig. 4.1), to produce MVA **4.6**, uses two moles of NADPH, and is catalysed by the enzyme HMG-CoA reductase.

All of these enzymes are now well characterized, and in most instances the segment of DNA coding for the enzymes has been identified and cloned. The most important enzyme is undoubtedly HMG-CoA reductase, since this catalyses the rate-determining step for the whole route to cholesterol. Inhibition of this enzyme thus provides a viable strategy for lowering levels of cholesterol in the blood, an important therapeutic goal since hypercholesteraemia is a recognized causative factor in heart disease.

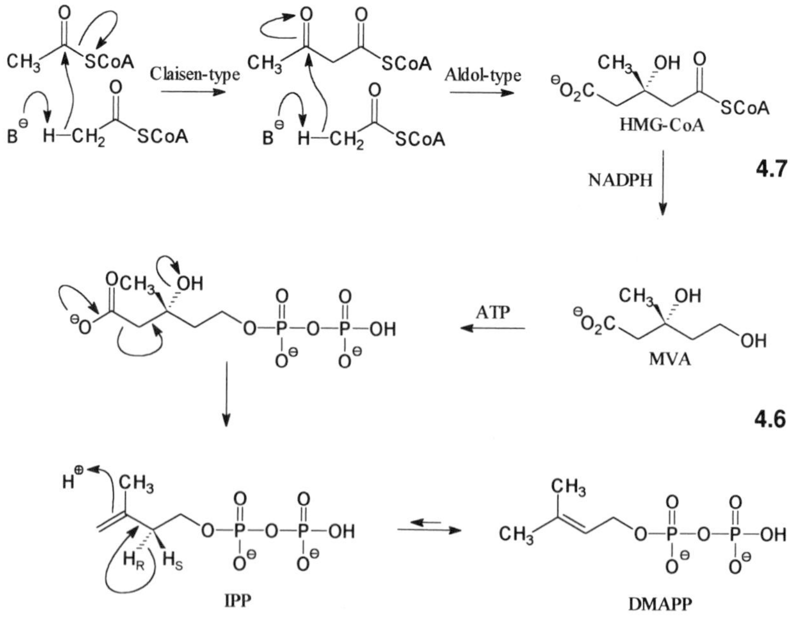

Fig. 4.1

Mevalonic acid is subsequently pyrophosphorylated to produce MVA-5-pyrophosphate, and this suffers decarboxylation to yield isopentenyl pyrophosphate (IPP), **4.10** (see Fig. 4.1), the first of the biogenetic isoprene units. A stereospecific isomerization then ensues to provide the other five-carbon unit, dimethylallyl pyrophosphate (DMAPP), **4.11** (see Fig. 4.1). The use of stereospecifically tritiated species of MVA was of pivotal importance in the elucidation of the mechanism of this and subsequent steps of the biosynthetic pathway. For example, note the stereospecific loss of the H_R hydrogen (a tritium in biosynthetic studies) during the isomerization of IPP to DMAPP.

4.2 Monoterpenoids

Combination of these two five-carbon units produces the monoterpenes, and it is now well established that prior ionization of DMAPP occurs to yield a tightly held ion-pair, which is highly electrophilic. This suffers nucleophilic attack by IPP to produce geranyl pyrophosphate, **4.12** (Fig. 4.2). Once again, the loss of hydrogen from IPP is completely stereospecific. The evidence for the formation of the ion-pair arises from studies using fluorine-containing analogues of DMAPP. For example, use of the trifluoromethyl analogue **4.13** (see Fig. 4.2) in place of DMAPP, as a substrate for enzymes (prenyltransferases) producing monoterpenes and higher isoprenoids, leads to rates of conversion that are typically one million times slower than the natural ones. This is reasonable if a carbocation is implicated, since the powerful negative inductive effect of the CF_3 group would destabilize such an ion, and thus reduce the rate of ionization. A bimolecular nucleophilic displacement reaction (see Fig. 4.2) would not involve a carbocationic intermediate, and would in consequence suffer no such retardation.

In 1976, the compound compactin, **4.8**, was isolated from the moulds *Penicillium citrinum* and *P. brevicompactum* and shown to possess marked inhibitory activity against HMG-CoA reductase. Two years later the structurally similar compound mevinolin, **4.9**, was isolated form *Aspergillus terreus*, and shown to be even more potent, and this natural product is in use in both the USA and Europe (under the trade name 'Lovastatin') for treatment of patients with dangerously high cholesterol levels.

4.8, R = H;

4.9, R = Me

4.12

4.13   **4.15**   **4.16**   **4.17**   **Fig. 4.2**

The geranyl pyrophosphate can then participate in two related processes. It can be a substrate for various cyclase enzymes which catalyse the formation of the cyclic monoterpenes, or it can act as a substrate for the prenyl transferase that catalyses the formation of farnesyl pyrophosphate, **4.14**, (see Fig. 4.5) the progenitor of the sesquiterpenes (C_{15} isoprenoids). The monoterpene cyclases have been intensively studied by Croteau and his group, and their results obtained with the enzymes from sage, *Salvia officinalis*, are good examples. Two separate cyclase enzymes have been identified, and geranyl pyrophosphate is first isomerized to yield a mixture of (-) (3R)- and (+) (3S)-linalyl pyrophosphates (Fig. 4.3). These then cyclize stereospecifically to provide (+)-α-pinene and (−)-β-pinene respectively. Use of enzyme inhibitors like the geranyl fluoride, **4.15**, and linalyl fluoride, **4.16**, which disfavour ionization, and the sulphonium analogue, **4.17**, which is a transition-state analogue, all support the intermediacy of the linalyl ion pair (shown in Fig. 4.3). (see Fig. 4.2 for structures **4.15–4.17**.)

All other cyclic monoterpenes can be considered to arise via rearrangement reactions that involve the initial cyclic species **4.18**, and a selection of probable pathways are shown in Fig. 4.4. Similar rearrangement processes can be executed in the laboratory, and some tracer experiments that have been carried out support the mechanisms shown in the figure.

Most of the mono- and bicyclic monoterpenes have characteristic aromas, and form the basis of the perfumery and flavour industry. Interestingly, several pairs of enantiomers have markedly different aromas. For example, (+)-carvone, **4.19**, smells of caraway seeds, whilst the (-) form has a spearmint aroma; (+)-limonene, **4.20**, smells of oranges whilst the (-)-form has the aroma of lemons. This provides clear evidence for the existence of olfactory receptors (in the nose) with a precise three-dimensional architecture.

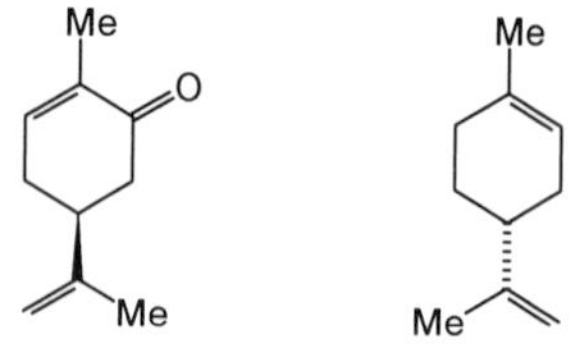

4.19   **4.20**

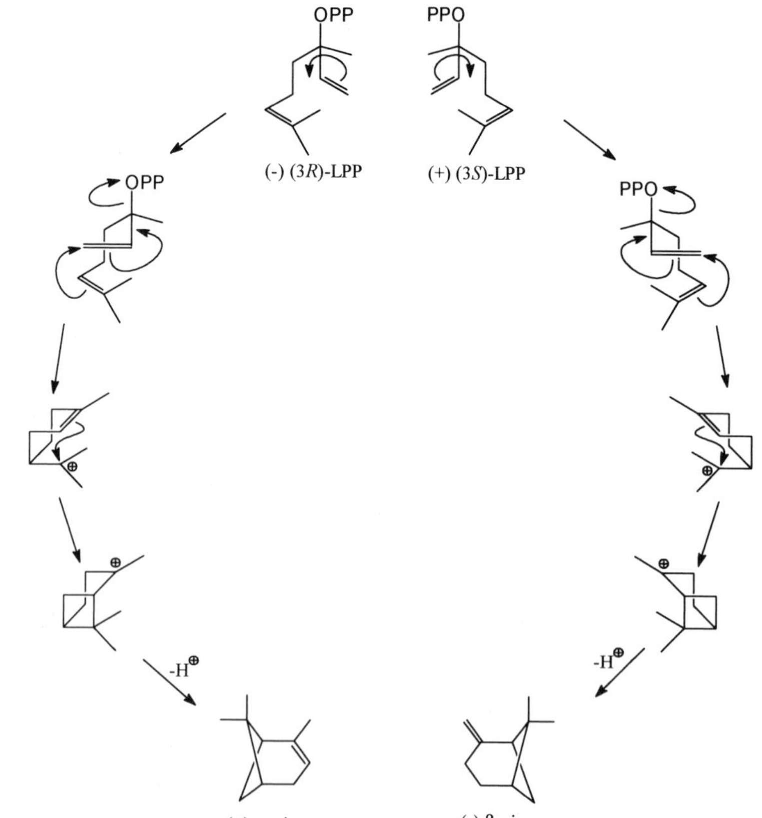

Fig. 4.3

(+) α-pinene   (-) β-pinene

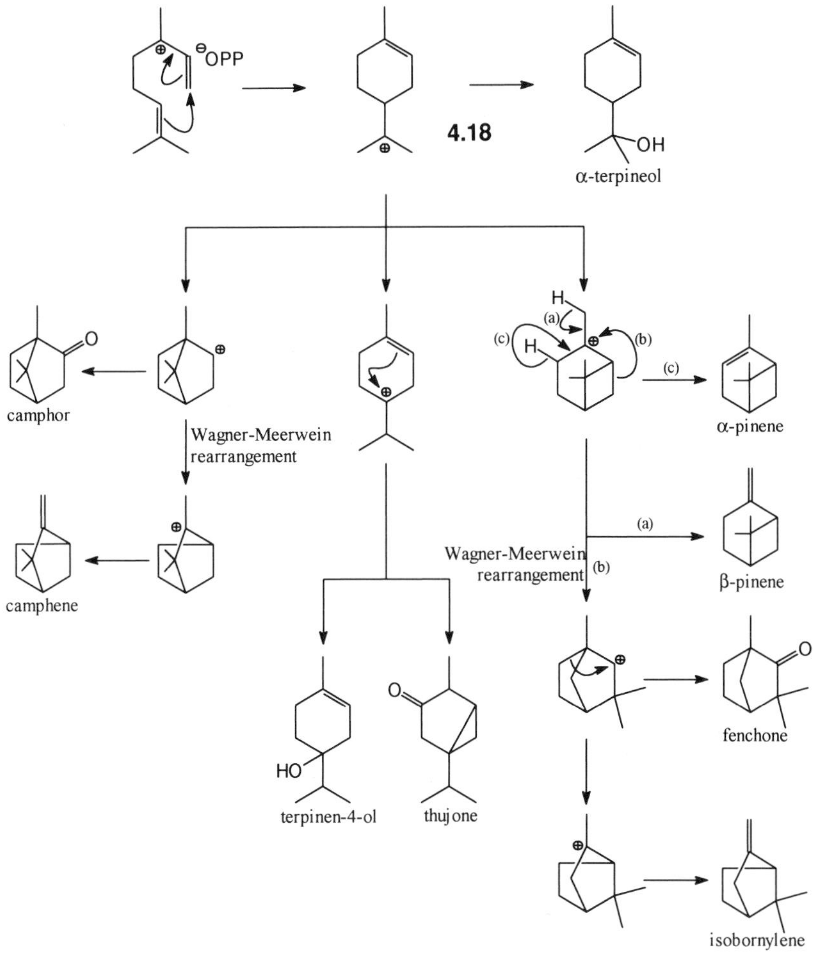

Fig. 4.4

Other monoterpenes have great historical interest and these include thujone (primarily (–)-3-isothujone), **4.21**, which was a major constituent of the liqueur absinthe, and ascaridole, **4.22**, which has been used for centuries as an anthelmintic, i.e. it kills intestinal parasites. Absinthe was a highly alcoholic extract of *Artemisia absinthium* and other shrubs, and was initially valued (by the Napoleonic armies) for its anthelmintic properties, though it later became infamous for its supposed aphrodisiac properties. Inveterate users included such literati and artisans as Lautrec, de Maupassant, van Gogh, Baudelaire, and Verlaine, all of whom extolled the virtues of the liqueur in their writings or paintings. Over a period of years, however, such absinthists all developed signs of mental derangement, and either died young or committed suicide. Not surprisingly, thujone has been cited as the culprit, and there is no doubt that it is neurotoxic, though all absinthists were undoubtedly alcoholics as well.

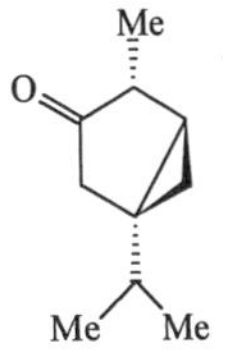

4.21

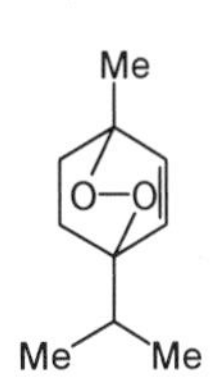

4.22

Ascaridole from the plant *Chenopodium ambrosioides* has been in use since at least the time of the Romans, and does possess marked anthelmintic properties. More recently it has been shown to have antimalarial activity as well, but unlike the sesquiterpene artemisinin, **4.3**, it has not been used clinically for this purpose.

4.3   Sesquiterpenoids

The formation of the basic sesquiterpenoid skeleton proceeds via the same mechanism as described for monoterpene formation. Thus geranyl pyrophosphate condenses with IPP to produce farnesyl pyrophosphate, **4.14** (Fig. 4.5), and this is the progenitor of all other sesquiterpenes. Most of these compounds are mono- or bicyclic in structure, and the various cyclase enzymes mostly utilize the isomer nerolidyl pyrophosphate, **4.23** (see Fig. 4.5), as the substrate for the various cyclization reactions. Some of these cyclases have been isolated and thoroughly investigated, and the biosynthetic pathway to trichodiene, **4.24** (Fig. 4.6), is exemplary. The whole process proceeds via a series of carbocation (ion-pair) intermediates, and involves the migration of a methyl group and a hydrogen. This pathway was originally suggested by Hanson and co-workers on the basis of tracer experiments, and has been substantiated by Cane and co-workers through the use of specifically labelled (with tritium and ^{13}C) substrates in conjunction with the purified enzyme, trichodiene synthase.

Fig. 4.5

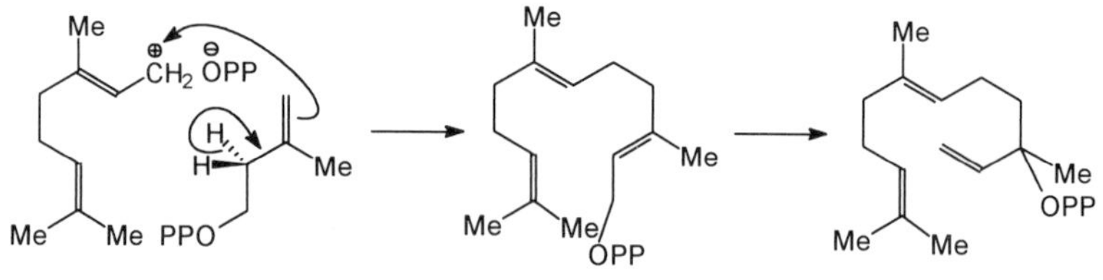

4.14   **4.23**

Trichodiene is of considerable interest because it is the precursor of the trichothecenes. These complex mycotoxins are produced by several species of *Fusarium*, and are common contaminants of grain, leading to chronic or acute illness in grazing animals and in humans. In one incident, during the winter of 1942–43, grain contaminated with trichothecenes caused the deaths of thousands of Russians. Due to the prevailing wartime conditions, the cereal had been left in the fields during the winter, then harvested and used in making bread, etc., which was heavily contaminated with *Fusarium* moulds. The resultant illness, termed alimentary toxic aleukia, characterized by persistent sore throat and lowering of the white blood cell count, lowered the immunity of the afflicted and made the susceptible to a range of life-threatening infections. Thousands died within a few weeks of consuming the grain products.

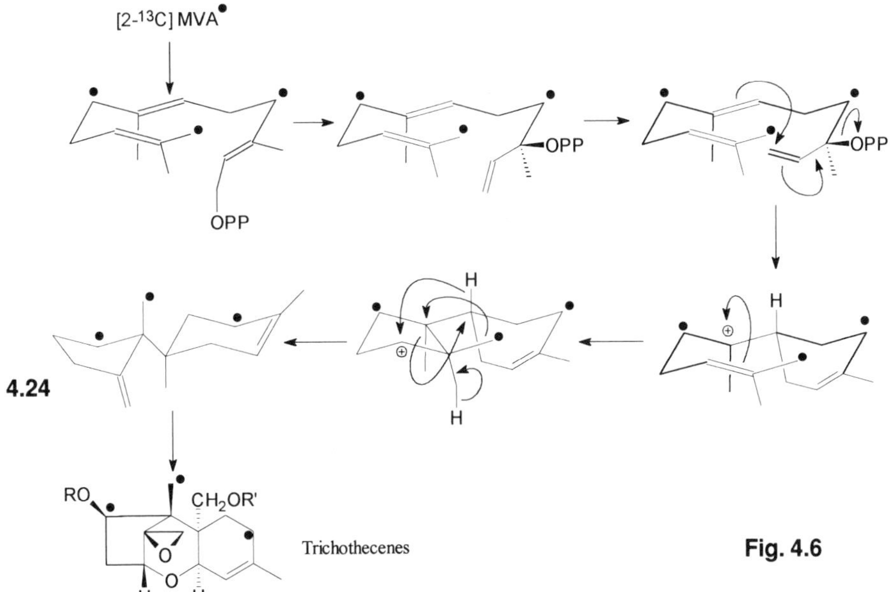

Trichothecenes

Fig. 4.6

The toxicity of these mycotoxins was invoked more recently as part of a cold-war 'squabble' between the USA and the USSR. In 1981 the Americans accused the Soviets of spraying parts of Vietnam with spores of *Fusarium*, citing as evidence a 'yellow rainstorm' that left foliage contaminated with trace amounts of trichothecenes. Thorough scientific investigation failed to substantiate these claims, and it is now certain that the phenomenon was due to mass aerial defaecation by swarms of the giant honey bee *Apis dorsata*, a not uncommon event in South–East Asia.

Another family of sesquiterpenes with important biological activities are the juvenile hormones (JH), e.g. **4.25** (see Fig. 4.7). These are essential hormones that control the growth and maturation of insects. In particular they must be present during the various larval stages, but absent if the insect is to metamorphose from larva (or chrysalis) into the adult insect. They were first isolated in 1965 from the abdomens of male silkworm moths, family *Cecropia*, and the structure elucidation was carried out on 300 µg of hormone using micro-ozonolysis coupled with the (then) newly introduced technique of gas-liquid chromatography/mass spectrometry.

At about the same time (1964), two biologists at Harvard, Williams and Law, were working with the larvae of the 'European bug' *Pyrrhocoris apterus*, and were surprised to discover that they would not metamorphose. Instead they underwent numerous larval moults to become giant larvae. Reasoning that the larvae were consuming the paper towels that lined the dishes in which they were kept, Williams and Law replaced these with newspapers from various countries. Interestingly, US papers produced the same results, whilst European and Japanese paper products allowed normal development and maturation of the larvae. Solvent extraction of the American paper towels provided the answer to the mystery—the sesquiterpene

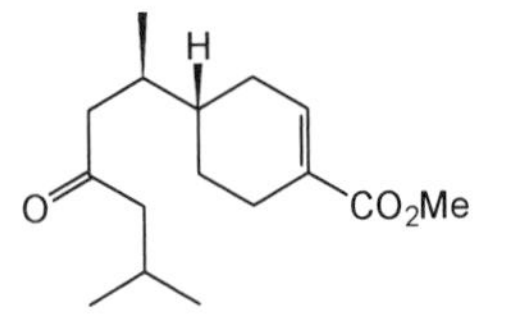

4.26

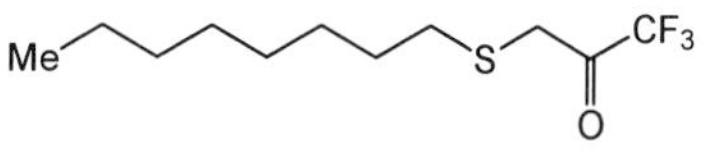

4.28

The thioketone **4.28** is a recent example of a compound that is a particularly potent inhibitor of the JH esterase. This enzyme assists with the metabolism of juvenile hormones, thus ensuring their removal prior to metamorphosis. The analogue **4.28** causes a 50 per cent inhibition of enzyme activity at a concentration of 1×10^{-9} molar, and promises to be of interest for both insect control and fundamental biological investigations.

juvabione, **4.26**. This is present in the bark of the American balsam fir tree, *Abies balsamea*, but not in the tree species used for European or Japanese paper production. Its structure has some similarities to that of the juvenile hormones, and it clearly acts as a juvenile hormone mimic, thus preventing the maturation of larvae growing on the species. It thus represents a further example of an ecological interaction between two species (tree and insect) mediated by means of a natural product.

The biosynthesis of the juvenile hormones, **4.25**, involves use of homomevalonic acid, **4.27** (see Fig. 4.7), in place of mevalonic acid as a progenitor of the six-carbon units (Fig. 4.7). There has been intense interest over the years in the potential of juvenile hormone analogues as insect-controlling agents.

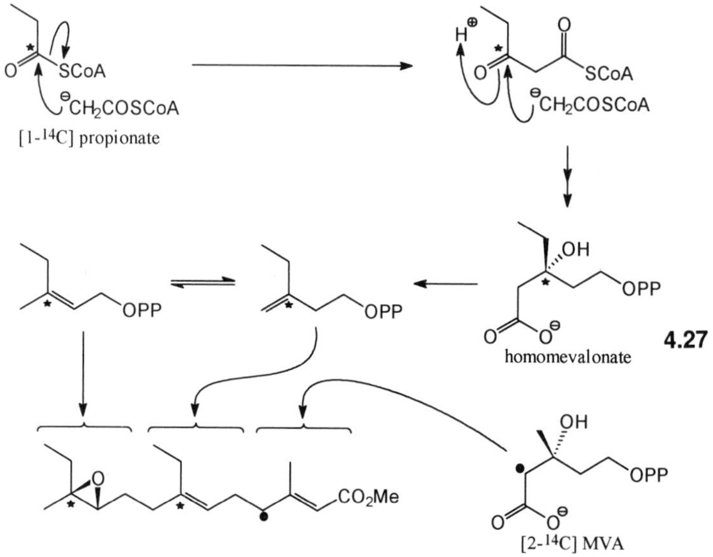

Fig. 4.7

4.25

One final sesquiterpene of special contemporary interest is artemisinin, **4.3**. This is produced by Chinese *Artemisia annua*, and extracts of the plant have been used for the prevention and treatment of malaria for at least 2000 years. In 1972, Chinese research workers isolated and proved the structure of artemisinin (or qinghaosu), and went on to demonstrate its biological potency. Clinical trials commenced in 1979, and in that year over 2000 patients suffering from malaria (caused by the parasite *Plasmodium falciparum*) were cured. The compound is also surprisingly non-toxic to mammals, and it is now a standard treatment in China. In the West, much effort has been directed towards the synthesis of analogues, and simplified structures like **4.29** retain most of the potency of artemisinin. Nothing is known of its biosynthesis, but Haynes has accomplished a rather neat, biomimetic synthesis commencing with the known sesquiterpene dihydroqinghao acid, **4.30** (Fig. 4.8).

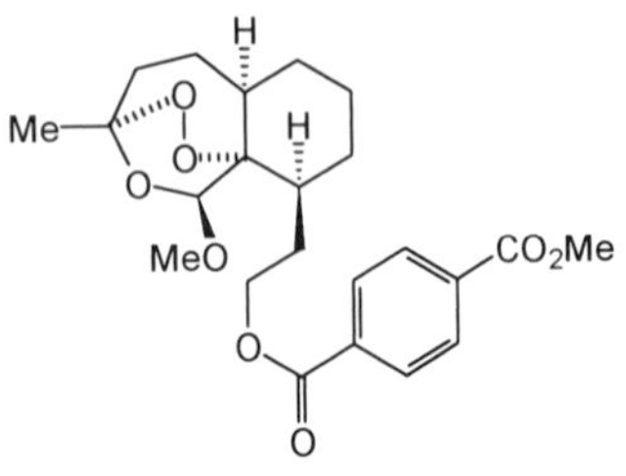

4.29

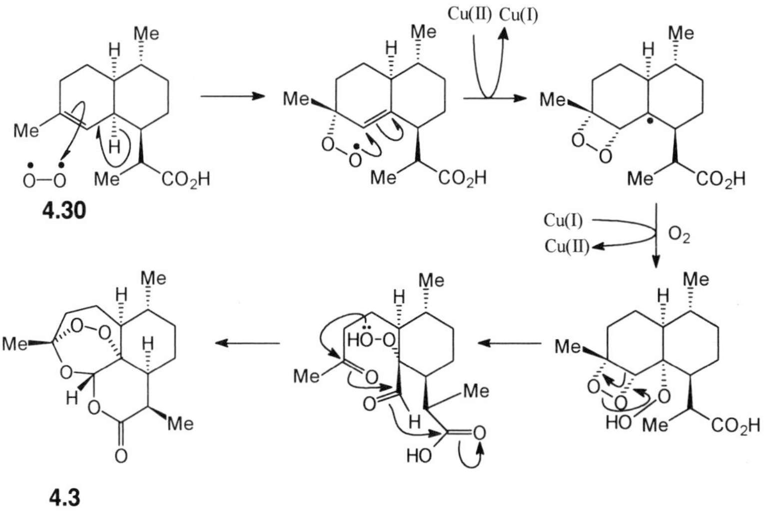

4.30

4.3

Fig. 4.8

4.4 Diterpenoids

The diterpenes are produced following the condensation of farnesyl pyrophosphate with IPP to produce gerenylgeranyl pyrophosphate, **4.31** (see Fig. 4.9), and thence the carbocation **4.32**, which is the progenitor of many of the diterpenes. The formal S_N2' reaction that is involved in the final stages of the cyclisation shown in Fig. 4.9 is of interest.

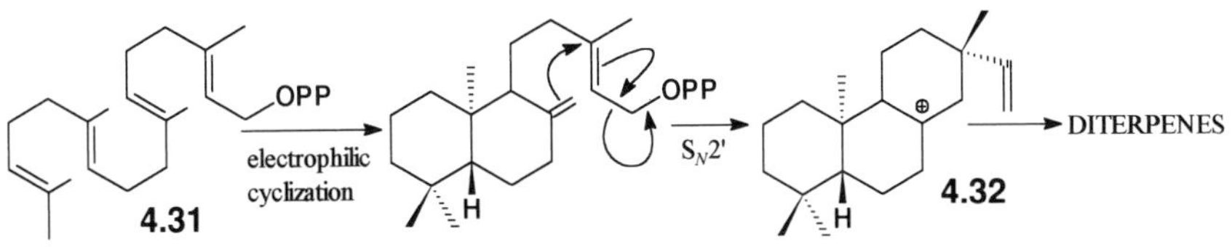

4.31 electrophilic cyclization S_N2' **4.32** DITERPENES

Fig. 4.9

Of all the diterpenoids, the gibberellins have received the most attention. These were first isolated from the fungus *Gibberella fujikuroi*, which is a pathogen that causes rapid elongation of rice seedlings (tall straggly plants are obtained). At least 60 discrete structures are now known, and many of these are believed to act as endogenous hormones for a range of plant species. Their biosynthesis is complex and only the key stages *en route* to gibberellin A$_3$ are shown in Fig. 4.10. Of particular note is the interesting rearrangement that occurs half-way through the sequence which may occur via the mechanism depicted.

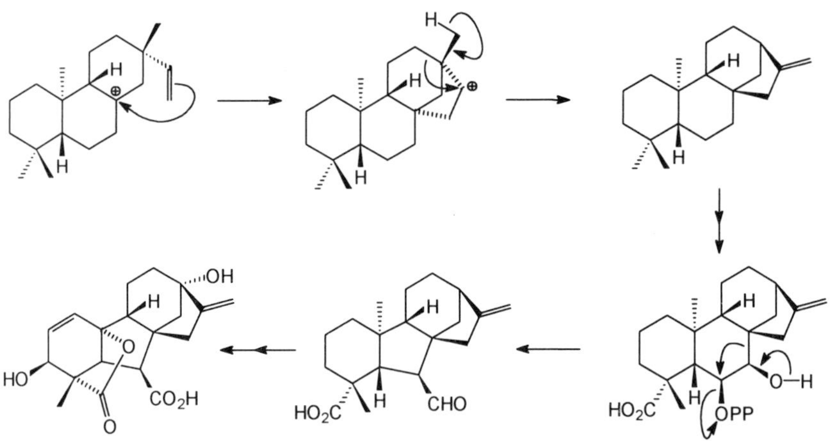

Fig. 4.10

gibberellin A$_3$

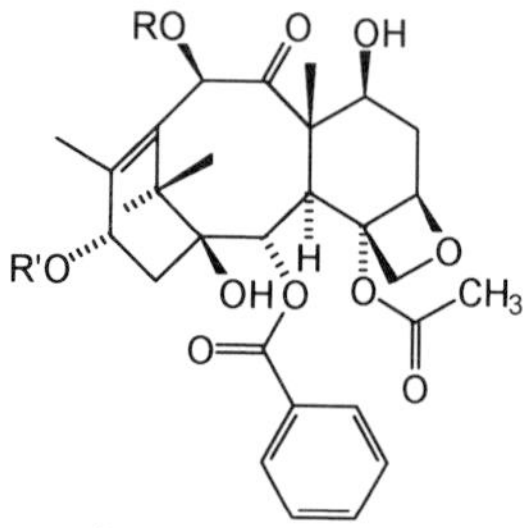

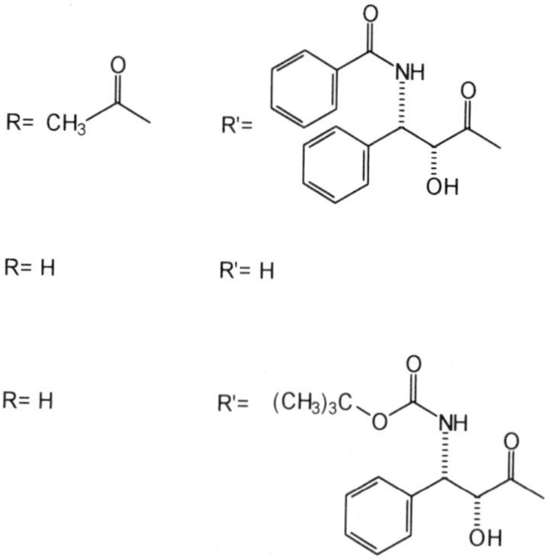

The highly complex diterpenoid taxol, **4.33**, is of great contemporary interest due to its extremely potent and broad spectrum antitumour activity. It is presently undergoing advanced clinical trials in the USA and Europe. In the past, most supplies of the compound have come from the bark of the Pacific yew, *Taxus brevifolia*, but since stripping the bark eventually kills the tree, this source is of limited utility. More recently, the related compound 10-deacetylbaccatin III, **4.34**, has been isolated in useful amounts from foliage of the common yew, *Taxus baccata*, and the derived semi-synthetic drug taxotere, is currently of great interest. The compounds enhance the polymerization of the protein tubulin, which is a central structural component of the cell, to produce stable microtubules. These have especially important functions during cell division, and the taxanes disrupt the normal equilibrium between soluble tubulin dimers and polymeric microtubules.

4.33   R= CH$_3$

4.34   R= H   R'= H

taxotere   R= H

4.5   Steroids and triterpenoids

The sesterterpenes (C$_{25}$ compounds) are formed via the same general mechanism, i.e. geranylgeranyl pyrophospate, **4.31**, condenses with IPP, but the triterpenoids and steroids are produced by a completely different pathway. The key to an understanding of their biosynthesis arose initially from the discovering of large amounts of the C$_{30}$ compound squalene, **4.35**

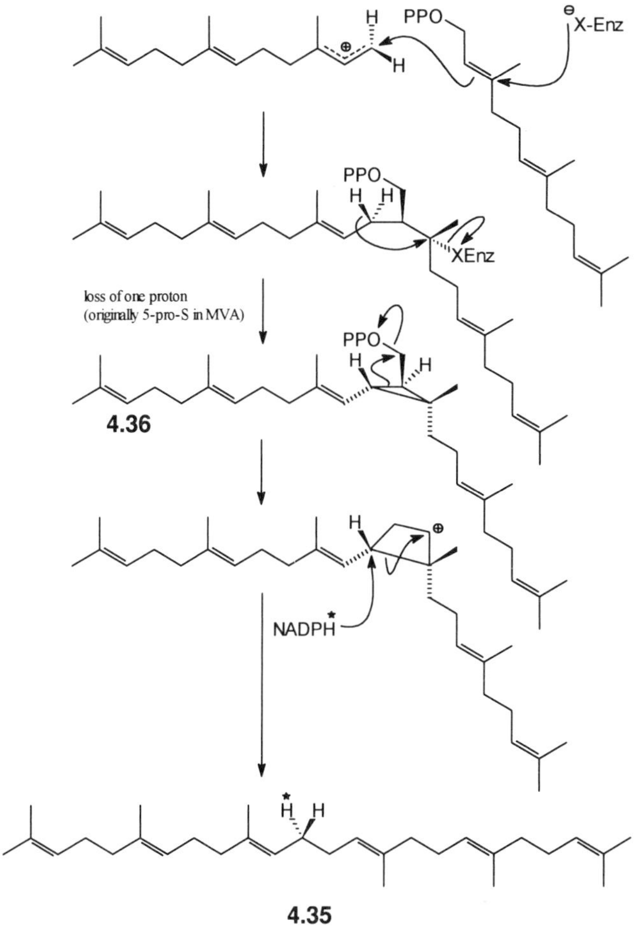

Fig. 4.11

(see Fig. 4.11), in shark liver oil. This structure clearly contained two far-nesyl moieties, but joined in a tail-to-tail fashion rather than in the head-to-tail fashion observed for all of the smaller terpenoids. The actual mechanism by which these two C_{15} units condense could not have been predicted, and actually proceeds via the formation of the cyclopropane, **4.36**—presqualene pyrophosphate (Fig. 4.11). Once again the importance of the electrophilic allyl cation is evident.

Squalene is then selectively epoxidized to form 2, 3-oxidosqualene (**4.37**; R = Me, X = CH_2; see Fig. 4.12) which then cyclizes in an apparently concerted fashion to yield the steroid lanosterol (**4.38**, Fig. 4.12). This fantastic sequence of events, catalysed by the two enzymes squalene epoxidase and epoxysqualene cyclase, has been rigorously investigated. The regiospecific epoxidation can be emulated in the laboratory, though it is then at best regioselective, and there is no doubt that squalene exists in a number of favoured, compact conformations like the one shown in Fig. 4.12. The terminal alkenes are thus exposed, and the epoxide acts as a latent electrophilic centre that is the focus of a series of electron migrations and ring closures to yield the protosterol **4.39** (R = Me, X = CH₂) or its equivalent (e.g. pyrophosphate ion-pair). A series of 1, 2-suprafacial shifts then occur to produce lanosterol.

Recent model studies by Corey and co-workers have done much to establish the structural and stereoelectronic requirements of the substrate, and to confirm the mechanistic details of the conversion. In particular, a protosterol **4.40** could be trapped when the epoxysqualene analogue **4.37** (X = O, R = Me) was employed; and the very strict conformational requirements were indicated when the *bis*-norepoxysqualene **4.37** (X = CH₂, R = H) was employed. The cyclase accepted this analogue as a substrate, but used it to produce the non-steroidal compound **4.41** (Fig. 4.12).

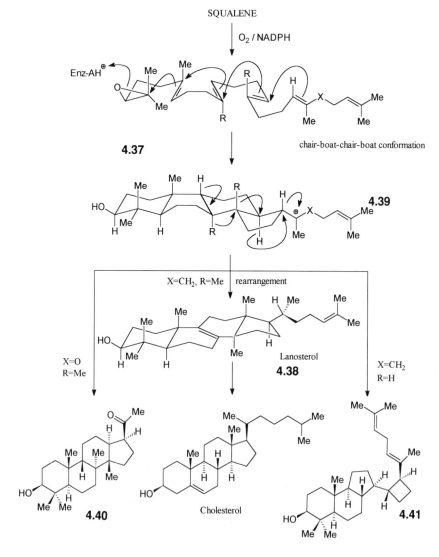

Fig. 4.12

Clearly this enzyme is a prime target for inhibition, if the production of steroids, especially cholesterol, is to be reduced. A number of squalene analogues do have marked inhibitory activity, for example **4.42**, though they are presently of more interest as experimental tools rather than as clinically useful drugs.

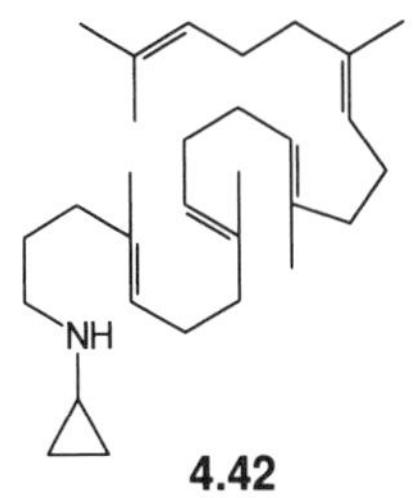

The conversion of dihydrolanosterol, into cholesterol, **4.4** (see Fig. 4.13), has also been studied intensively, and proceeds via a series of oxidative demethylations catalysed by enzymes of the cytochrome P_{450} family. These employ a haem cofactor whose role is to activate molecular oxygen, which then attacks the methyl groups converting them in turn into hydroxymethyl, formyl, and thence via loss of methanoic acid, to the demethylated product (Fig. 4.13). Most of the intermediates shown in this figure have now been isolated.

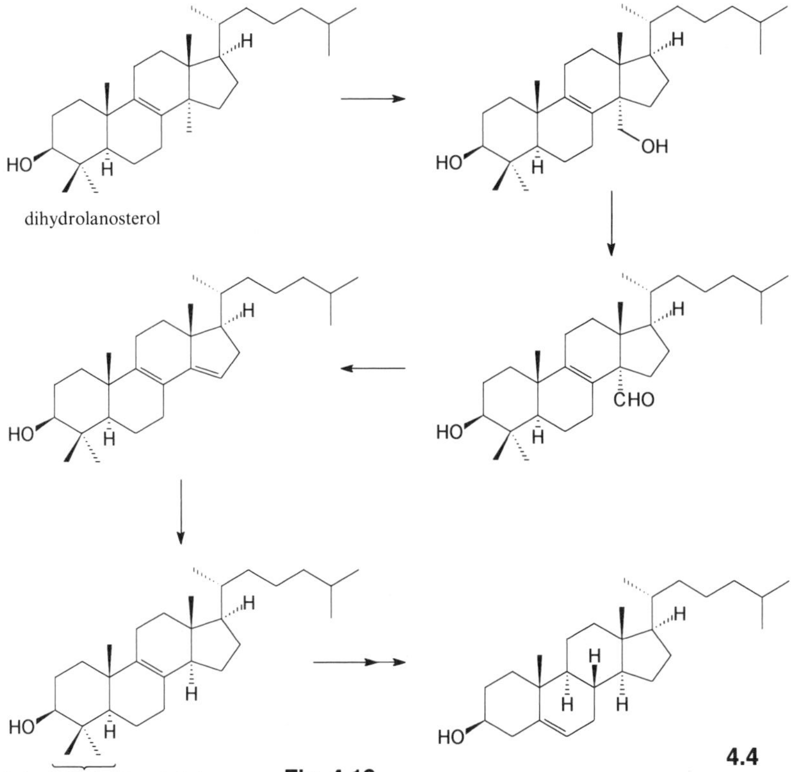

4.42

dihydrolanosterol

4.4

further oxidative demethylation   **Fig. 4.13**

The biosynthetic pathway shown in Fig. 4.13 is restricted to mammals, and the phytosterols found in plants and yeasts are produced via the pathway shown in Fig. 4.14. This proceeds via the cyclopropyl intermediate cycloartenol, **4.43**, but similar 1, 2-shifts and oxidative demethylations occur. Some of these phytosterols also contain extra carbon atoms in the side chain, and these are added via the sequence shown in Fig. 4.15. In this way the C_{27} precursor is converted into the C_{28} and C_{29} phytosterols like ergosterol, **4.44**, and sitosterol, **4.45**.

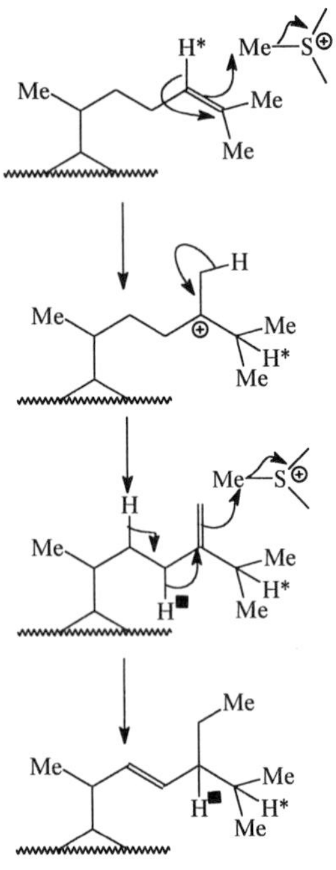

Fig. 4.15

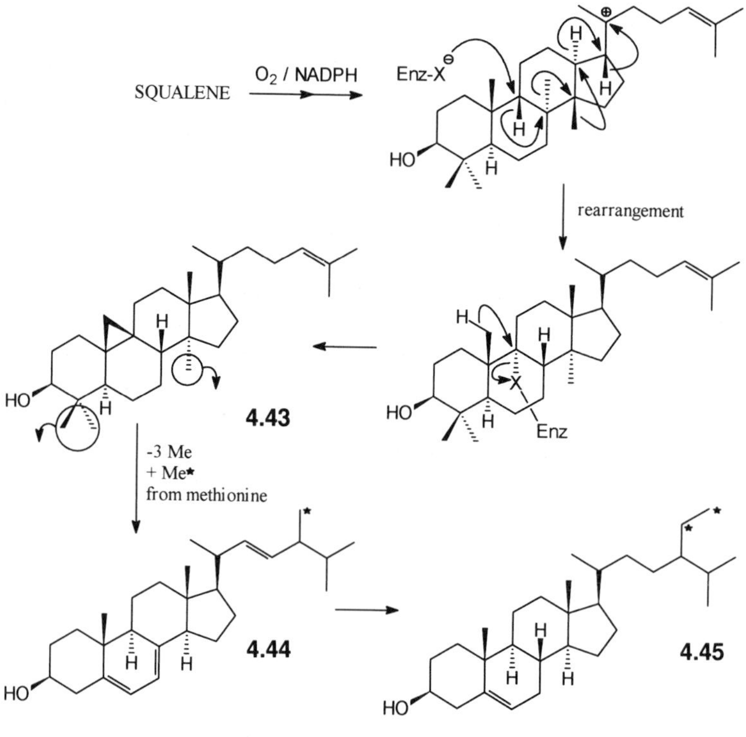

Fig. 4.14

Most plants also contain triterpenoids (C_{30}), and these are also derived from 2, 3-oxidosqualene, but a slightly different chair–chair–chair–boat conformation is involved. The biosynthesis of β-amyrin, **4.46** (Fig. 4.16), is typical.

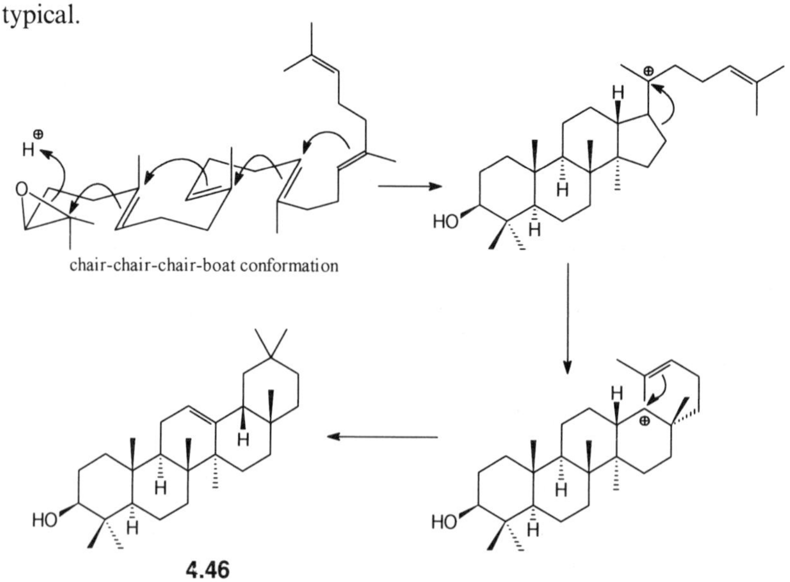

chair-chair-chair-boat conformation

Fig. 4.16   **4.46**

Biological function of steroids and triterpenoids

When considering the biological functions of these molecules, we are confronted with the fact that on a total weight basis, many of the compounds are relatively long-lived, i.e. they are not metabolized to any extent. They thus appear to have an essential role as constituents of cell membranes where they help to maintain structural integrity and control permeability.

All eukaryotes (organisms that contain a discrete cell nucleus) synthesize steroids, or have an absolute requirement for them in their diet. Even certain prokaryotes (organisms without a discrete cell nucleus) have the capacity to effect a cyclization of squalene to form steroid-like or triterpenoid-like molecules. Thus *Tetrahymena pyriformis* can convert squalene into tetrahymanol, **4.47**, and this compound probably has a vital structural role in the organism. Lower organisms often use hydroxylated carotenoids or acyclic polyprenols for the same function; and certain bacteria make hopanoids, e.g. **4.48**, which act as membrane stabilizers. It has been estimated that the latter class of compounds contributes around five per cent of the total soluble organic matter of sediments.

4.47   **4.48**

Other steroids and triterpenoids are metabolized to produce biologically active molecules like the human sex hormones, vitamins D, and the insect moulting hormones, and these are of more obvious interest. It is generally accepted that insects cannot produce their own cholesterol (or similar basic steroid), but use steroids present in their diet as starting materials for the construction of compounds necessary for membrane functions or as hormones. The moulting hormones, e.g. ecdysone **4.49**, are produced by most insects and (in conjunction with the juvenile hormones) control the various stages of moulting prior to metamorphosis. Their biosynthesis from cholesterol has been investigated extensively, but the only definite intermediate thus far identified is the unsaturated ketone **4.50**. It is interesting to note that crustaceans, i.e. crabs, lobsters, crayfish, etc., also produce moulting hormones and these have obvious structural similarities to the insect hormones

4.49, R = H
4.51, R = OH

4.50

(e.g callinecdysone-B, **4.51**). The occurrence of similar metabolites and probably similar biosynthetic pathways within the insects and crustaceans is not surprising since they evolved from the same primeval proto-insect/crustacean creatures.

The biosynthesis of the mammalian sex hormones has also been investigated thoroughly, and the pathway from cholesterol to progesterone, **4.52**, and thence to androstendione, **4.53**, testosterone, **4.54**, and oestrone, **4.55**, is well established (Fig. 4.17). Most of the enzymes involved are cytochromes of the P_{450} family, and like the demethylases of cholesterol biosynthesis, they employ a haem cofactor to activate oxygen which then serves as a source of peroxy anion or hydroxy radicals. The proposed biogenesis of the female sex hormone oestrone, which is catalysed by the enzyme aromatase (Fig. 4.17), is representative.

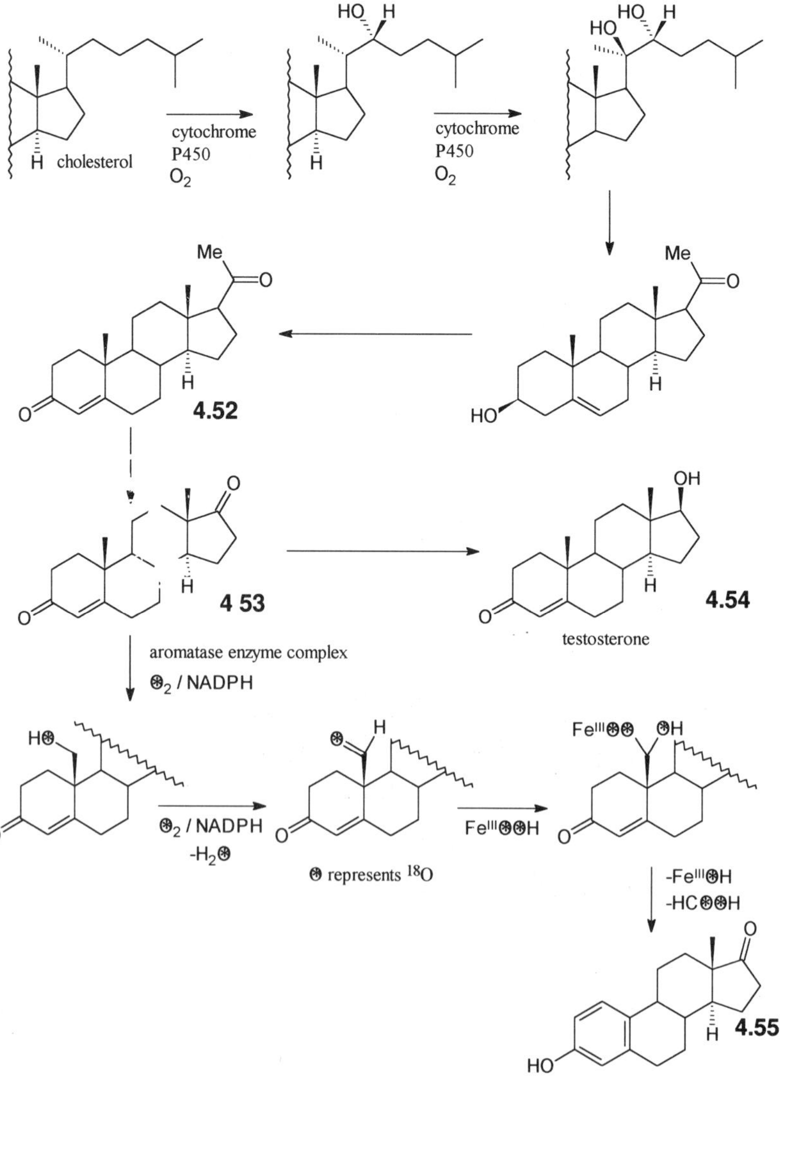

Fig. 4.17

About one third of mammary tumours require a supply of oestrogens to support their growth, so inhibition of aromatase is an obvious target for chemotherapy. A number of drugs have been designed, and many of these are mechanism-based inhibitors, that is they inhibit the enzyme by subverting the chemistry used for the conversion of androstendione to oestrone, or compete for binding sites on the enzyme. Two examples are shown in Fig. 4.18. In the upper example the drug acts as a mechanism-based inhibitor (a kind of suicide substrate), in that the enzyme converts it into an acyl fluoride via the usual oxidative mechanism, and this then acylates the enzyme thus causing inhibition. In the second example, the drug has a functional group that allows it to coordinate to the haem iron, thus occupying the coordination site that would normally be occupied by oxygen. No activation of the oxygen can then take place.

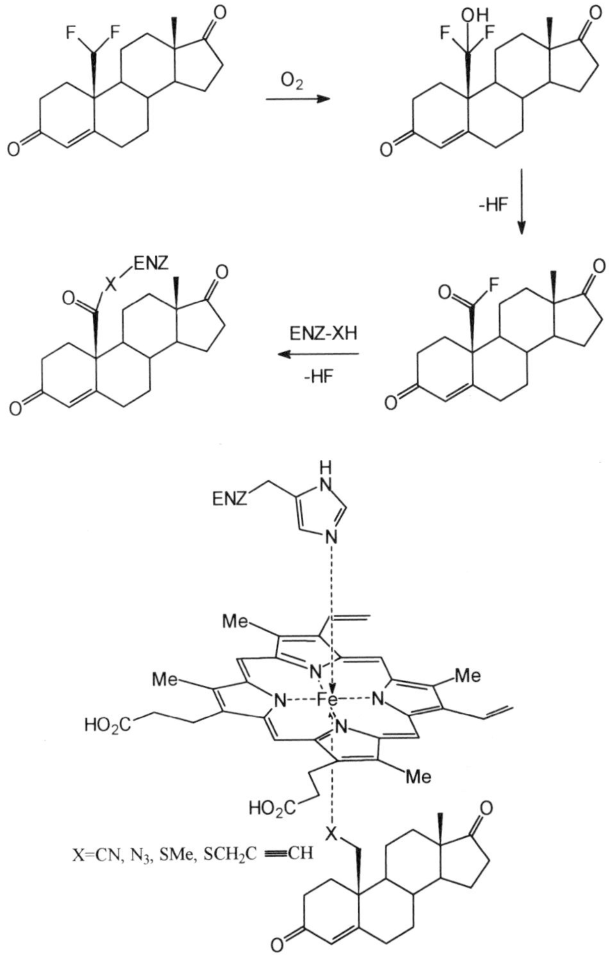

Fig. 4.18

The female sex hormones, oestrogens and progesterone, and male sex hormones, most importantly testosterone, control the development of secondary sexual characteristics (e.g. breast development and beard growth, respectively) during puberty, and the oestrogens are also intimately involved in the control of the menstrual cycle. During the cycle the ovaries come under the influence of not only the steroid hormones but also two peptide hormones (from the pituitary) known as lutenizing hormone (LH) and follicle stimulating hormone (FSH). These initially stimulate the growth of the ovum-containing follicle. The oestrogens are primarily responsible for maturation of the ovum, whilst progesterone is produced around the time of ovulation and then helps to prepare the uterus for reception of the fertilized ovum. It is subsequently produced throughout pregnancy by the placenta, and helps to maintain the pregnant state by inhibiting production of LH and FSH (so-called 'feedback inhibition'). This ensures that the menstrual cycle is temporarily halted.

An understanding of these complex chemical interactions had to await the isolation and structure elucidation of the sex hormones, which occurred during the period 1925–35. The pharmaceutical companies were now excited by the prospect of having access to quantities of these compounds for the control of fertility and problems caused by imbalance of the hormones. The synthesis of these steroids was accomplished during the 1930s and 1940s, but they remained extremely expensive. This situation changed almost overnight with the discovery by Russell Marker that diosgenin, **4.56**, from the Mexican yam (family *Dioscorea*) could be converted in four simple chemical steps into progesterone, **4.52**. The pharmaceutical company Syntex was created to capitalize on this discovery, and they were soon able to supply progesterone at 48 cents a gram compared with the earlier price of $80 a gram for synthetic material. The modern contraceptive steroids like the progestin norethindrone, **4.57**, arose from this revolutionary discovery.

4.56   **4.57**

This mimics the activity of progesterone, thus inhibiting the production of FSH and LH, and causing temporary infertility. Other therapeutically useful steroids also had their origins at this time, and the highly effective anti-inflammatory steroids, which are analogues of the natural anti-inflammatory steroid cortisone, **4.58**; betamethasone valerate, **4.59** (Betnovate, for topical use); and beclomethasone dipropionate, **4.60** (Becotide, for use as an inhaled antiasthmatic drug) are representative examples.

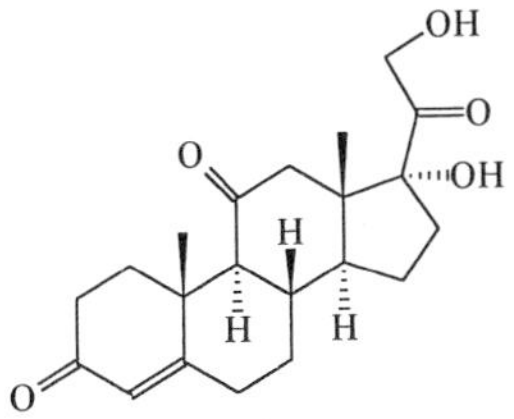

4.58 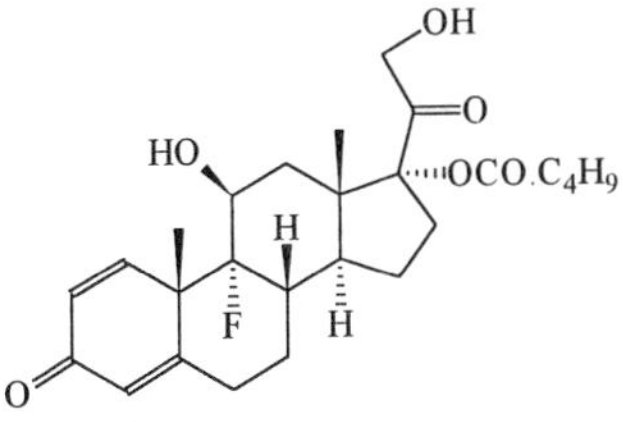   **4.59** 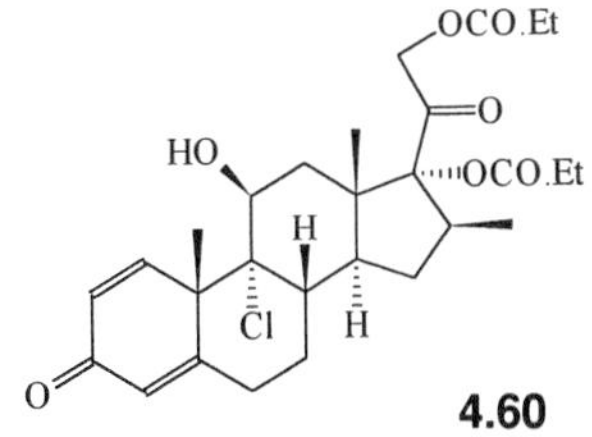   **4.60**

Cholesterol also acts as precursor of the bile acids, **4.61**, vitamin D₃, **4.62**a (see Fig. 4.19) (in mammals), whilst ergosterol, **4.44** (Fig. 4.14), acts as a precursor of vitamin D₂, **4.62**b (in yeasts).

The biosynthetic pathway to the vitamins D is illustrated in Fig. 4.19, and provides an example of a six-electron photochemically allowed ring-cleavage. Cholesterol, produced in the liver, is transported via the blood vessels to the fine blood capilliaries of the body surface, where it is converted into dehydrocholesterol, **4.63**, and the photochemical reaction then occurs. A rationalization of the stereochemical requirements of this process was one of the major achievements of Woodward and Hoffmann's rules for the conservation of orbital symmetry.

Cod liver oil, a rich source of vitamin D₃, was formerly a folk remedy in Northern Europe for the treatment of tuberculosis and rheumatism, but it gained prominence as a very effective treatment and prophylaxis for rickets in the 1860s. Rickets in children, and osteomalacia in adults, are due to a lack of calcium in the diet or its inefficient assimilation from the diet. This results in deformation of the growing bones or brittleness in older bones. Prior to the industrial revolution rickets seems to have been relatively uncommon in Europe, since most of the inhabitants worked out of doors and

The bile acids (or rather their sodium salts) are major constituents of bile, and the only ones which are active in digestion. They emulsify fats that are present in the part-digested food mass, producing tiny droplets of fat which are more susceptible to enzymatic cleavage. They are biosynthesized from cholesterol in the liver, and together with the other constituents of bile, are stored in the gall bladder.

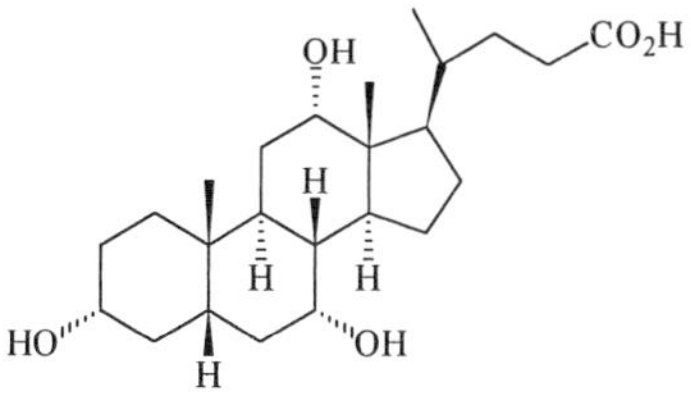

4.61

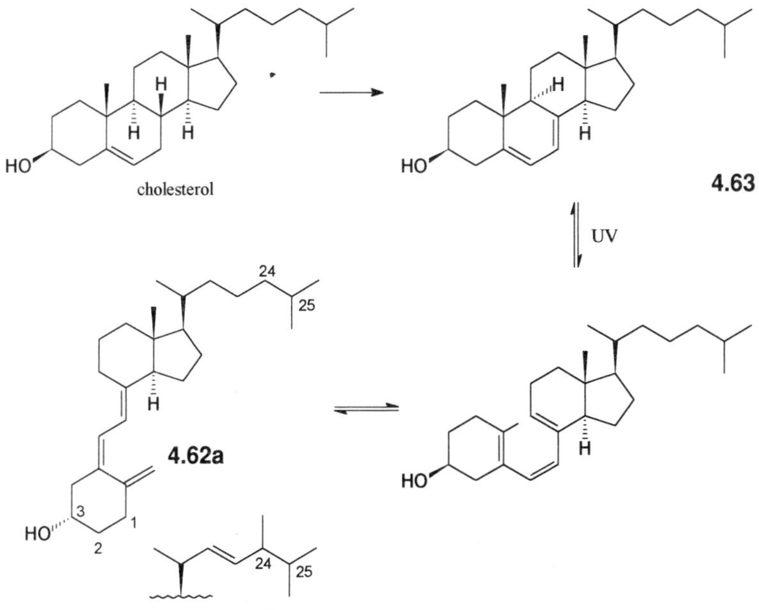

Fig. 4.19

hence received an adequate amount of UV exposure. Once the revolution was underway, the cities of Northern Europe became blanketed by a pall of smoke and other pollutants. People moving into the cities to find work were thus denied access to UV light, and they ceased producing enough vitamin D_3 to ensure efficient assimilation of calcium from their diet. These selfsame diets were also woefully deficient in calcium and vitamin D_2 from plant sources. In consequence, rickets and other diseases of poor nutrition became rife.

Although cod liver oil was first used on a large scale in the 1860s, it was not until the pioneering work by Windaus and others on the structure and chemistry of the vitamins that the link with UV light was appreciated. And it was not until the mid-1970s that the biochemistry of the vitamins D was understood. Vitamin D_3 is not in fact the active metabolite used for calcium assimilation, but is converted in the kidneys into several hydroxylated metabolites, most importantly 1, 25-dihydroxy-D_3 and 1α, 24, 25-trihydroxy-D_3. These then assist with the assimilation of calcium from the diet. An excess of calcium in the bloodstream inhibits the 1-hydroxylase of the kidneys, and this acts as one of the control mechanisms of calcium balance in the body.

4.6   Carotenoids

The carotenoids are, with few exceptions, C_{40} polyolefinic metabolites, and are common constituents of green plants, and certain algae, bacteria, and fungi. They lack the complex three-dimensional structures of the other classes of isoprenoids, and most of them have the all-*trans* arrangement of their conjugated double bond systems. Many of them undoubtedly act as accessory pigments (to chlorophyll) in photosynthesis, and β-carotene, **4.5** is the most important in this respect.

The core biosynthetic pathway is well established, and proceeds via prephytoene pyrophosphate, **4.64** (compare this with the intermediacy of presqualene pyrophosphate); *cis* and *trans*-phytoene, **4.65**; and lycopene **4.66**; via a series of *trans*-eliminations of hydrogen, and thence by cyclization to yield β-carotene (Fig. 4.20).

4.67

4.68

4.69,   CHO
4.70,   CH$_2$OH

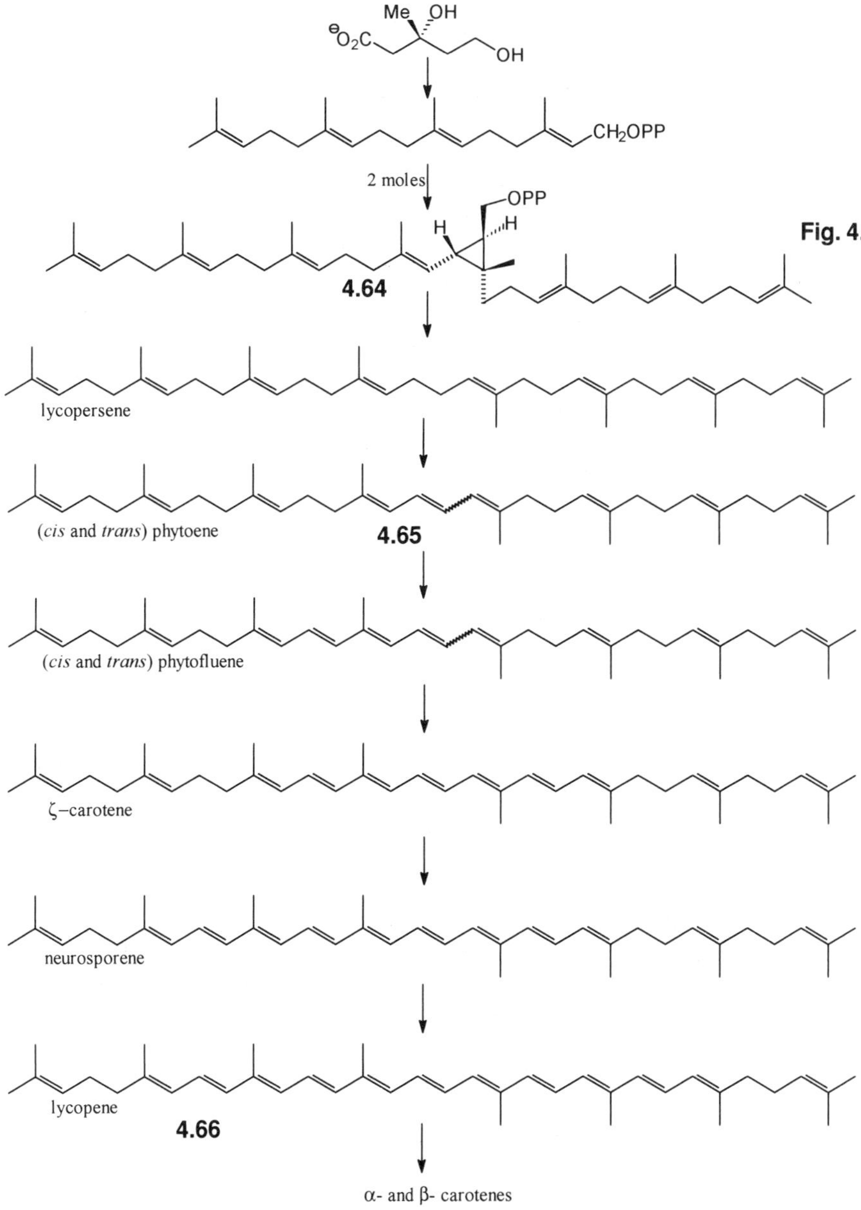

Fig. 4.20

Further oxidative metabolism yields xanthophylls like violoxanthin, **4.67**, and capsanthin, **4.68**, the pigment of red peppers. Oxidative degradation produces retinal, **4.69** (a major participant in vision processes) and thence vitamin A$_1$, **4.70**, and also the plant hormone abscicic acid, **4.71**, which controls bud development and seed germination.

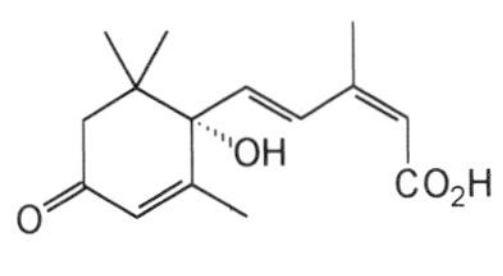

4.71

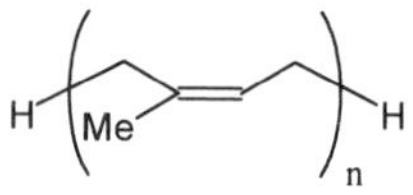

n = 500-5000

4.72

About one per cent of all plant species have the capacity to synthesize *cis*-polyisoprenoids, and the most important of these is rubber, **4.72**. The commercial source of rubber, *Hevea brasilensis*, has been commercially selected for its ability to convert MVA almost exclusively into rubber.

4.7   Products of mixed metabolism

A number of important secondary metabolites are produced with the help of two or more metabolic pathways, and the psychoactive constituents of *Cannabis sativa*, for example tetrahydrocannabinol, **4.73** (see Fig. 4.21), are prime examples. Although the actual pathway is unknown, the intermediacy of the polyketide metabolite olivetol, **4.74** and geranyl pyrophosphate (or an equivalent) are most likely (Fig. 4.21). The psychoactive properties of marijuana (hemp, pot, charas, bhang, etc.) were probably discovered in the twelfth century, and prior to that the plant was of value as a source of protein and oil (in the seeds) and hempen rope (the fibrous stem of mature plants).

The isoprenoid pathway contributes molecular fragments to numerous other secondary metabolites, and these will be discussed in later chapters.

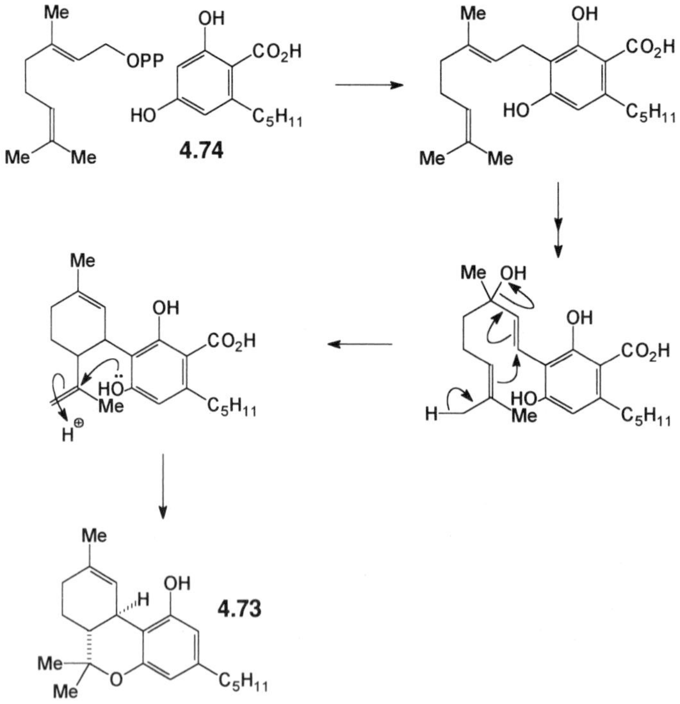

Fig. 4.21

5 The shikimic acid pathway: biosynthesis of phenols, lignans, flavonoids, etc.

Shikimic acid, **5.1** (see Fig. 5.1) is of pivotal importance in metabolism since it is the progenitor of the aromatic amino acids phenylalanine, **5.2**, tyrosine, **5.3**, and tryptophan, **5.4** (see Figs 5.6 and 5.7). These, in turn, are used for the construction of peptides and proteins, but also act as precursors of some alkaloids. The discovery of the importance of shikimic acid arose from the studies of mutants of the bacterium *Escherichia coli*. Normal strains of the bacterium can produce all three aromatic amino acids from glucose as a sole carbon source. However, certain mutant strains that were obtained following exposure of the parent bacterium to ionizing radiation would only grow in the presence of all three amino acids and *para*-aminobenzoic acid and *para*-hydroxybenzoic acid. Clearly the mutant bacteria possessed modified genes that could not code for the production of the enzymes needed to convert glucose into the key acids. Eventually it was discovered that shikimic acid could act as a replacement for glucose in all these mutants, and was thus a key intermediate *en route* to aryl-C_1, aryl-C_2 and aryl-C_3 compounds.

5.2

5.3

5.4

5.1 Biosynthesis of shikimic acid and the aromatic amino acids

The biosynthetic pathway to shikimic acid is well established, and is shown in Figs 5.1 and 5.2. Intriguing chemistry is involved at almost every stage of this biosynthesis. In the first step, phosphoenol pyruvate, **5.5** (a participant in the process of glycolysis, see Fig. 1.2) reacts with the four-carbon sugar erythrose-4-phosphate, **5.6**, to produce 3-deoxy-D-arabino-heptulosonate-7-phosphate (DAHP), **5.7**. This is formally an aldol reaction, and probably proceeds via an initial nucleophilic attack at phosphorus.

The second step is the most interesting, and has received much attention. If we write DAHP in its hemi-ketal form, **5.8**, the mechanism suggested by Knowles is easier to follow (Fig. 5.2). Oxidation of **5.8** to the ketone **5.9** (note the use of the cofactor couple $NAD^+/NADH$ for oxidation, i.e. removal of a pair of hydrogens) then allows the phosphate group to act as an internal base to facilitate its own elimination. Reduction of the ketone (NADH/NAD$^+$ mediates the addition of a pair of hydrogens) is followed by a second proton abstraction. This initiates ring-opening, and the rest of the sequence is believed to proceed spontaneously to produce 3-dehydroquinate, **5.10**.

The use of isotopically-labelled enol pyruvates showed that C-O bond cleavage rather than P-O bond cleavage was involved during the aldol reaction to form **5.7**. Hence the mechanism shown below and in Fig. 5.1.

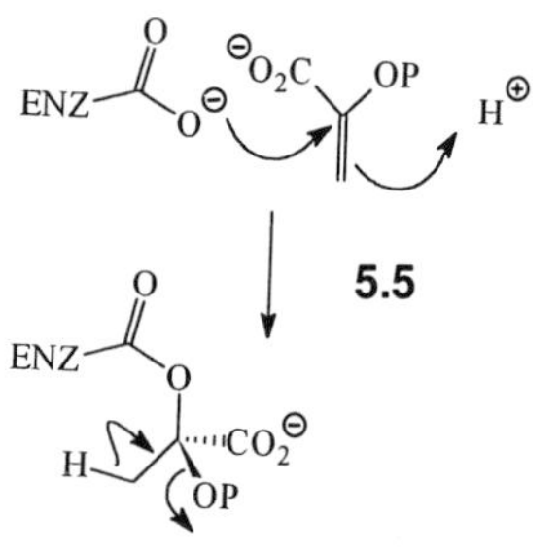

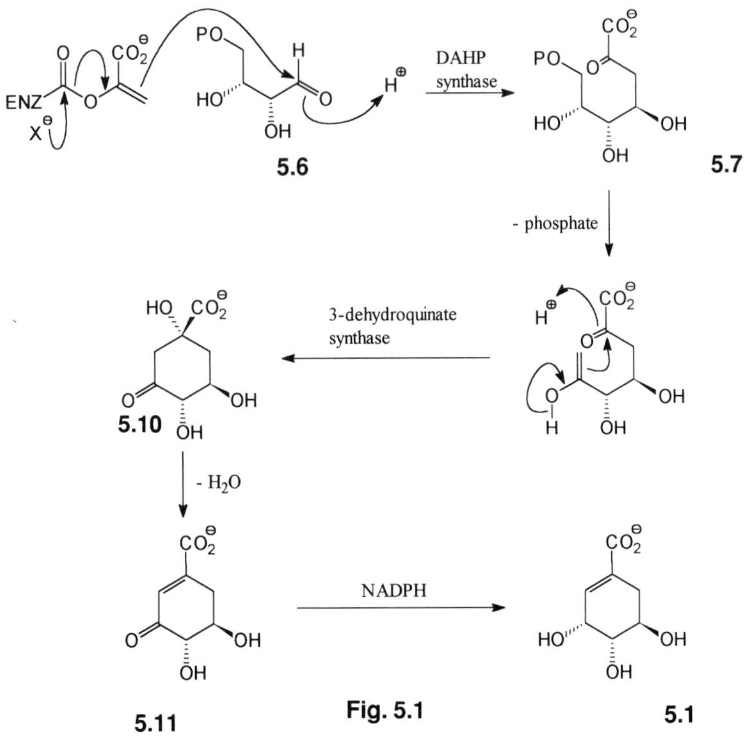

Fig. 5.1

The subsequent formation of 3-dehydroshikimate, **5.11**, proceeds via an equally intriguing mechanism, since it involves a *syn*-elimination of water. As this is unlikely to be a concerted process, the mechanism shown in Fig. 5.3 has been suggested. Initial condensation with the enzyme forms the Schiff's base **5.12**, and this then suffers loss of a proton followed by loss of hydroxide, and finally hydrolysis of the Schiff's base. Reduction of 3-dehydroshikimate then provides shikimate (**5.1**, carboxylate form).

The conversion of shikimate into the various amino acids is no less mechanistically interesting. The initial step involves formation of shikimate-3-phosphate followed by transesterification of shikimate with phosphoenol pyruvate to yield the enol ether **5.14**. This proceeds via the tetrahedral intermediate **5.13** (Fig. 5.4), and this has been isolated and characterized. Under the influence of the enzyme chorismate synthase, this enol ether then undergoes loss of phosphoric acid to produce chorismate, **5.15**.

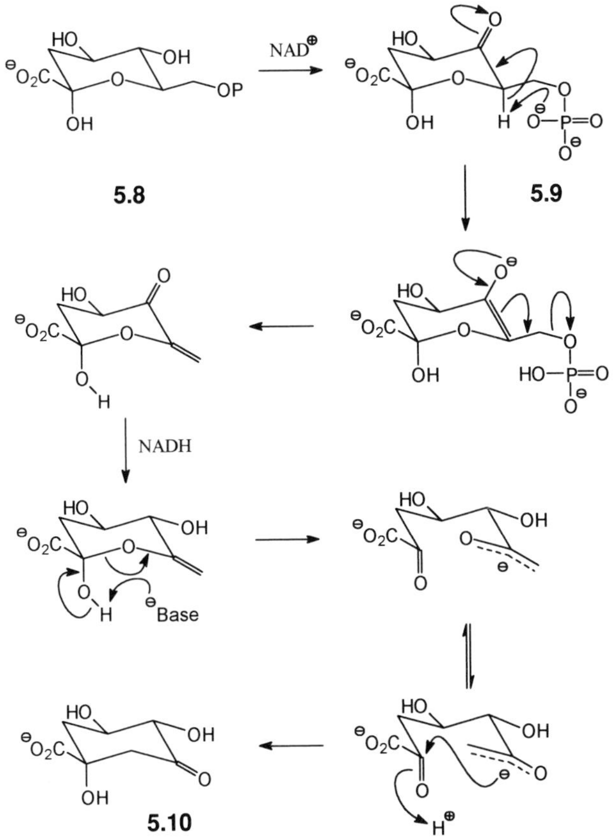

Fig. 5.2

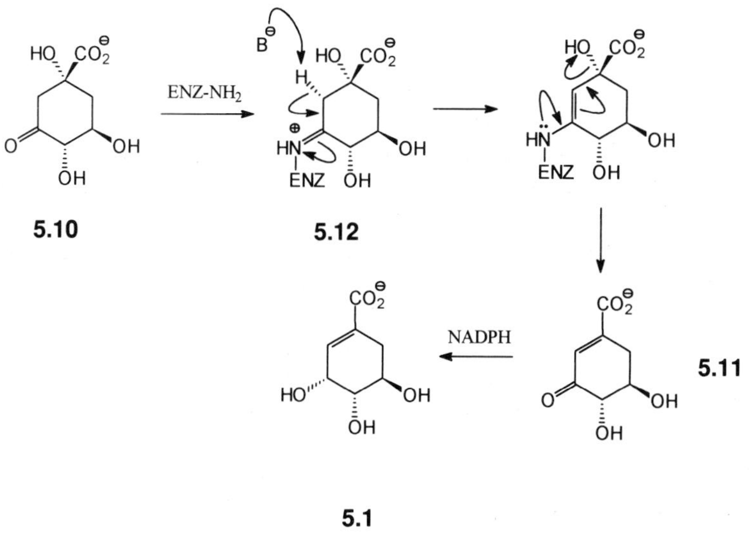

Fig. 5.3

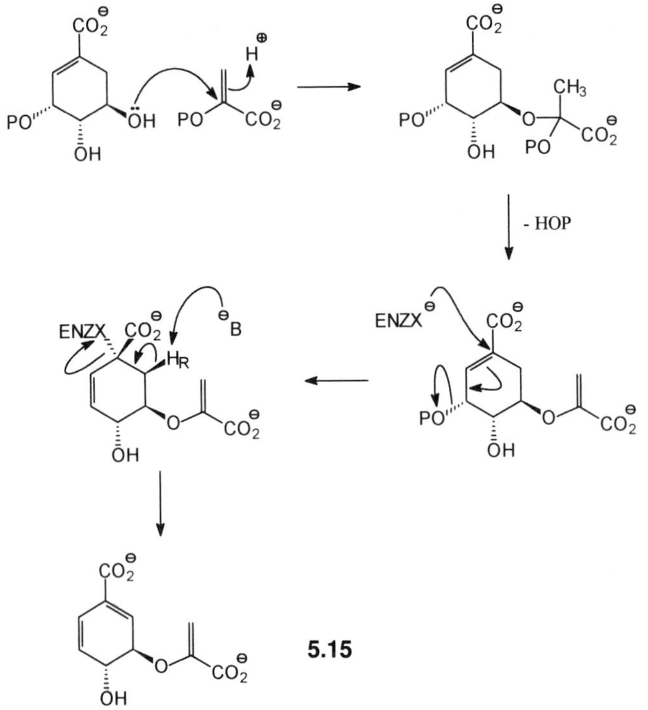

5.13

5.14

5.15

Fig. 5.4

A concerted *anti*-1, 4-elimination is chemically unfavourable, and the experimental data suggest that an initial covalent complex is formed between the enzyme and the substrate (Fig. 5.4). Certainly the stereospecific loss of H_R is not rate-determining, since the kinetic isotope effect of H_R in comparison with D_R was identical. However, an alternative mechanism, in which the phosphate leaves first to generate a carbocation, is also a possibility.

The second step, catalysed by the much studied enzyme chorismate mutase, involves the only known natural example of a Claisen ether rearrangement (Fig. 5.5). A number of potent enzyme inhibitors have been prepared, e.g. **5.16** and **5.17**, and these mimic the proposed transition state **5.18**.

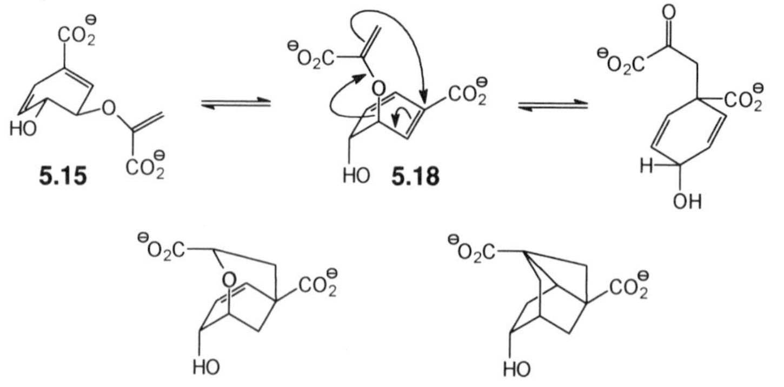

5.15 5.18 5.19

5.16 5.17

Fig. 5.5

Finally, various pathways exist for the transformation of prephenate, **5.19**, into phenylalanine **5.2** and tyrosine **5.3**, and these are summarized in Fig. 5.6. The actual pathway used depends upon the organism.

Tryptophan **5.4** is derived from chorismate, **5.15**, via initial formation of anthranilate, **5.20** (Fig. 5.7). The addition of ammonia probably involves a double S_N2' sequence. Anthranilate is coupled with ribose and then via indole-3-glycerol phosphate, **5.21**, produces tryptophan, **5.4**.

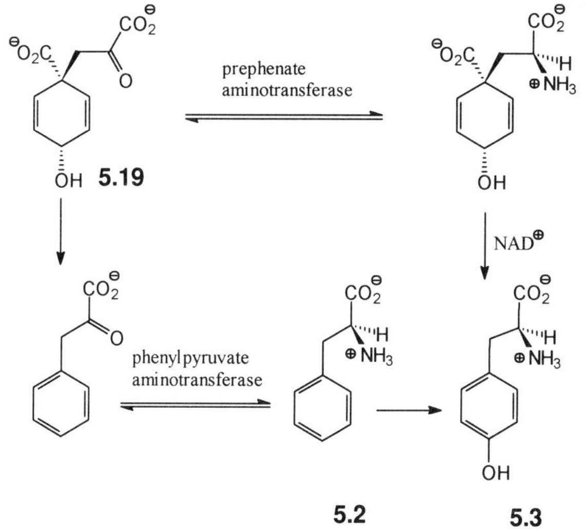

Fig. 5.6

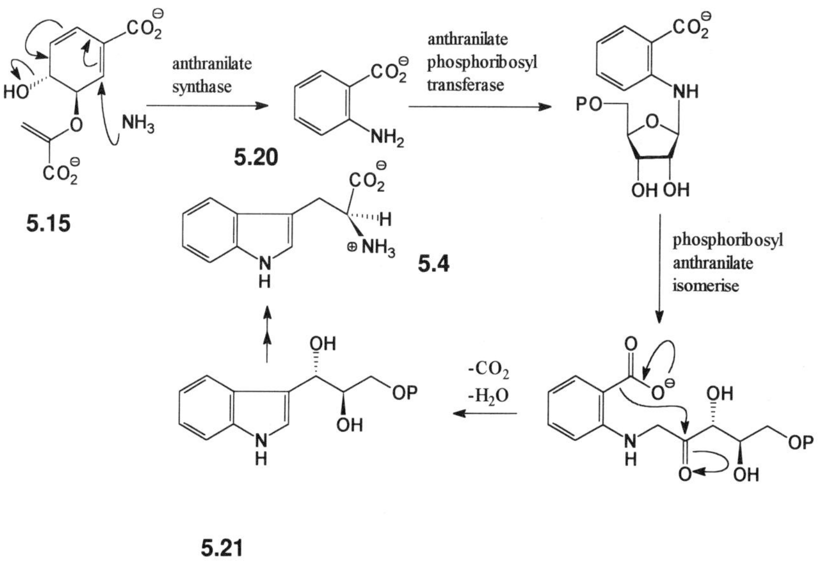

Fig. 5.7

Lignans, like podophyllotoxin, **5.23**, are common constituents of many plants, and often have interesting pharmacological properties. The Mayflower pilgrims discovered that the Indians of new England used roots of the indigenous plant *Podophyllum peltatum* to cure warts. Indeed, podophyllum resin is still available from pharmacies to treat warts. Investigation of the chemistry and pharmacology of these root extracts in the 1940s led to the isolation of podophyllotoxin, and the demonstration that it was a powerful inhibitor of mitosis (cell division). It inhibits the enzyme tubulin polymerase, which is required for the production of the protein tubulin. This is the major constituent of the fibres that hold together the two daughter cells during cell division. Inhibitors of mitosis have obvious utility in the treatment of cancer, and the semi-synthetic derivative of podophyllotoxin, etoposide (**5.24**) is in clinical use as a very effective treatment for testicular teratoma and small cell lung cancer.

5.2   Lignans, flavonoids, and other metabolites

Further metabolism of phenylalanine (and to a lesser extent tyrosine) produces cinnamic acid, **5.22** (or 4-hydroxycinnamic acid, see Fig. 5.8). The enzyme concerned is phenylalanine ammonia lyase (PAL), and catalyses a *trans*-elimination of ammonia. Cinnamic acid (and its 4-hydroxy analogue) are the precursors of most aryl-C_3 and (by degradation) aryl-C_1 and aryl-C_2 compounds, and a few representative examples are shown in Fig. 5.8.

The biosynthesis of podophyllotoxin probably proceeds via the route shown in Fig. 5.8 which involves stereospecific enzyme-mediated coupling of the two cinnamyl alcohols **5.25**. Non specific free radical coupling of various cinnamyl alcohols gives rise to polymeric structures (lignins), which represent around 50 per cent of the structural material of most woody plants.

Many of the other important shikimate metabolites contain structural components produced by other biosynthetic pathways. For example, the furanocoumarins like psoralen, **5.26**, and isopimpinellin, **5.27**, include a two-carbon unit derived from dimethyallyl pyrophosphate (and hence from MVA—see Chapter 4). The co-occurrence of marmesin, **5.28**, provides good evidence for the proposed biogenesis shown in Fig. 5.9.

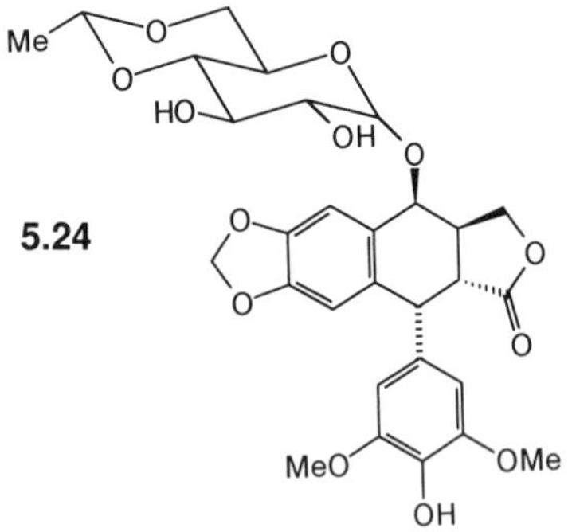

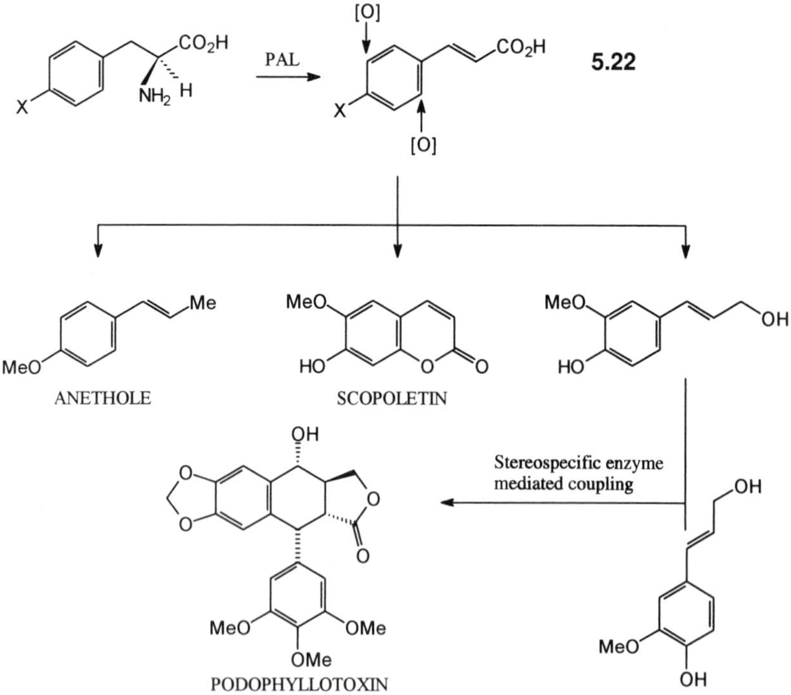

Fig. 5.8

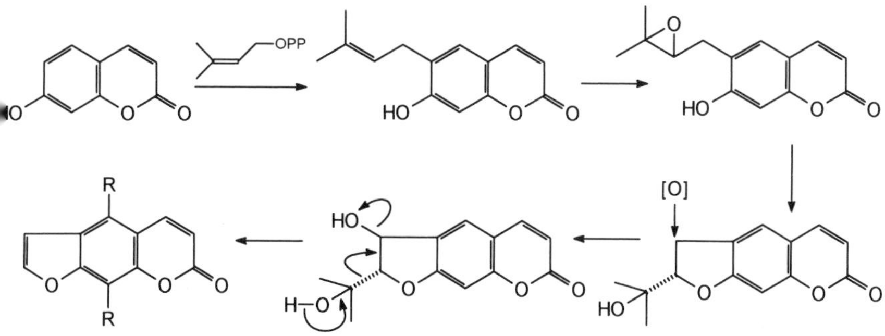

5.28

5.26, R = H

5.27, R = OMe

Fig. 5.9

The flavonoids comprise a large family of secondary metabolites that derive part of their structures from shikimate and part from the polyketide pathway (Fig. 5.10). They are usually found as conjugates with various carbohydrates (i.e. glycosides), and provide much of the colour of flowers and insect wings. Their biosynthesis proceeds via the condensation of the

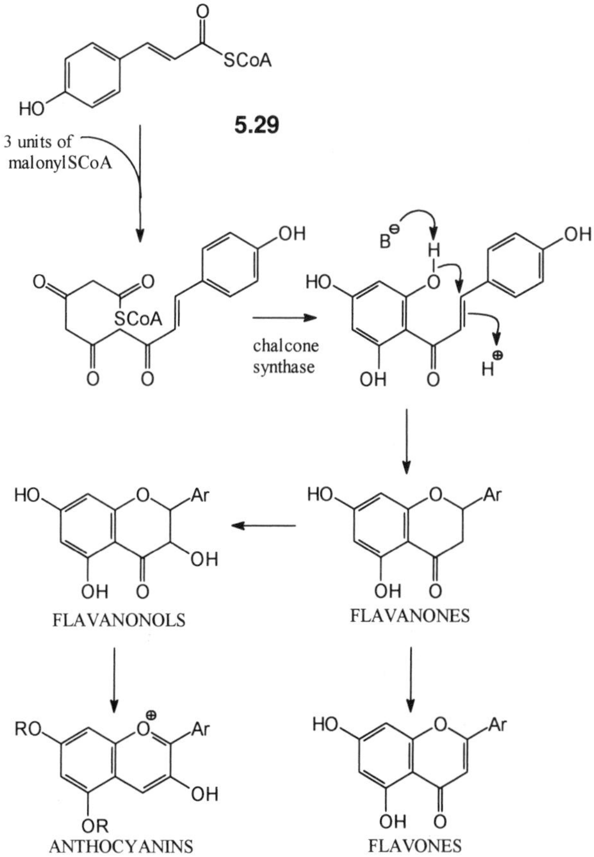

5.29

5.30

5.31

Fig. 5.10

coenzyme A ester of 4-hydroxycinnamic acid, **5.29**, and a triketide, a reaction catalysed by the enzyme chalcone synthase. Chalcone, **5.30**, then cyclizes to form the archetypal flavanone structure, **5.31**, a process catalysed by the enzyme chalcone isomerase. Note the different oxygenation patterns of the aromatic rings derived from the polyketide pathway (*meta*-disposition) and the shikimate pathway. Most other classes of flavonoids then arise via simple modification to the structure of **5.31**, though the isoflavonoids are probably produced via the somewhat more complex pathway shown in Fig. 5.11. As usual, an hydroxylase of the cytochrome P_{450} family is involved in this process.

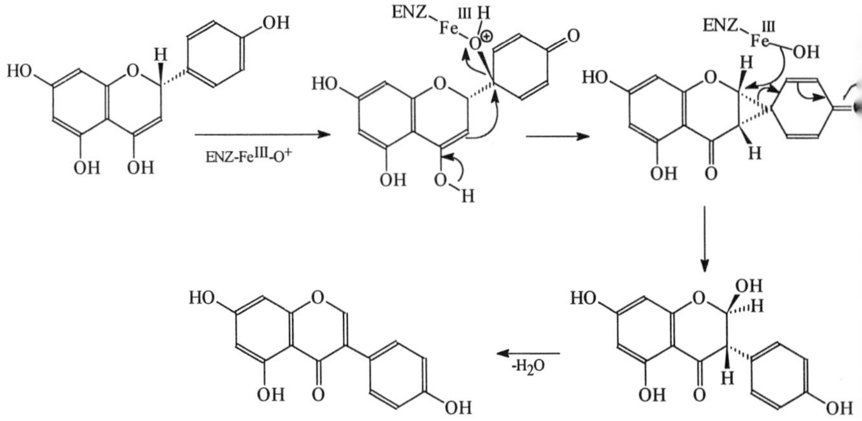

Fig. 5.11

As well as their importance as pigments (especially the anthocyanins), the flavonoids also contribute to the flavour and astringency of plants. For example, naringin (**5.32**) from grapefruit peel is intensely bitter, and the polymeric flavonoids like the condensed tannin structure **5.33** are very astringent. These compounds undoubtedly contribute to the 'defensive chemistry' of plants, since bitterness and astringency can be detected by most grazing mammals, including man.

It is interesting to speculate upon the origins of the flavonoids as a family. They are not found in marine plants, and since the compounds are good absorbers of UV light, it is probable that they 'evolved' when the marine plants first colonized the land. They thus served to screen the plants from harmful UV irradiation until large amounts of oxygen had been produced, which in turn gave rise to the UV-screening ozone layer.

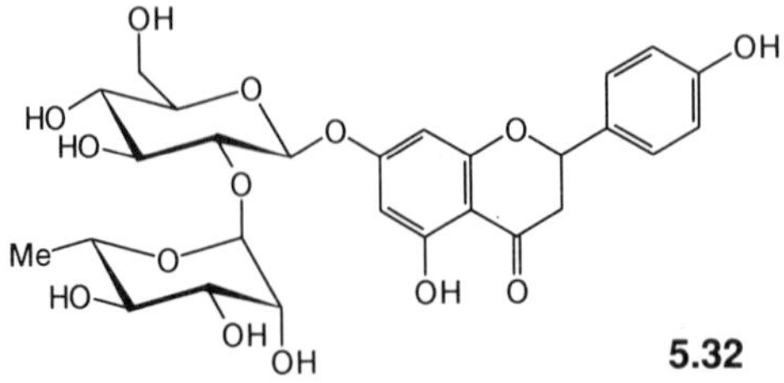

5.32 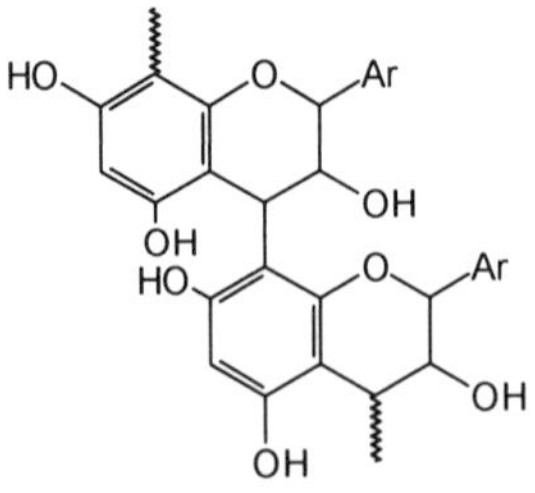   **5.33**

Finally, a number of diverse structures are also formed from chorismate, **5.15** (Fig. 5.12). Chloramphenicol, **5.34**, is produced from 3-(*p*-amino-phenyl)-(*S*)-alanine, **5.35**, whilst the vitamins K, e.g. **5.36**, are produced via isochorismate, **5.37** (see Fig. 5.13). (Note the addition of water to choris-mate (**5.15**) via a double S_N2' sequence.)

Chloramphenicol was first isolated in 1947 from an actinomycete (subsequently called *Streptomyces venezuelae*) isolated from a soil sample originating from Caracas in Venezuela. It proved to have broad-spectrum antibacterial activity, and was especially effective for the treatment of typhus, which had hitherto been untreatable. Between 1949 and 1951 an estimated eight million patients were treated with the drug, and as a result the American company Parke Davis, manufacturers of the drug, became, for a time, the biggest pharmaceutical company in the World in terms of turnover. However, the discovery that the drug produced fatal side-effects in a small percentage of patients, and the advent of the penicillins, put paid to this supremacy, and chloramphenicol is now largely used for the treatment of eye and ear infections.

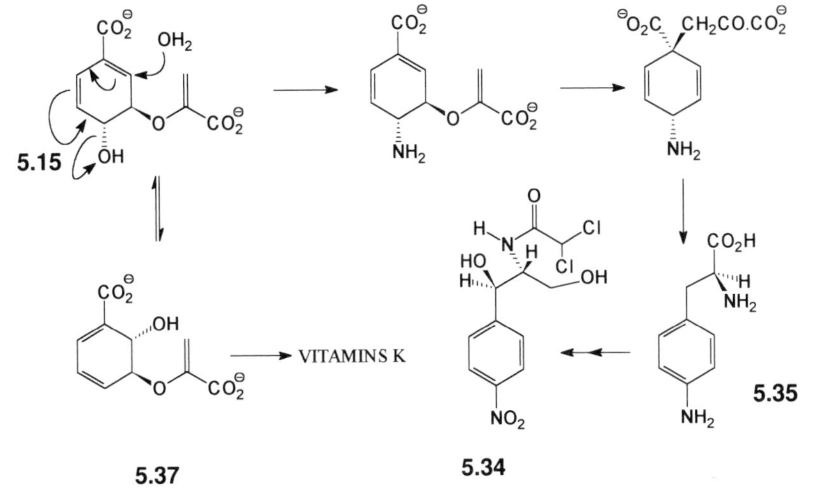

Fig. 5.12

The K vitamins are produced by bacteria residing within mammalian intestines (i.e. by the gut flora), and are required as essential cofactors for the production of prothrombin and other plasma proteins that are involved in the coagulation of blood. Their biosynthesis involves conversion of iso-chorismate **5.37** into *ortho*-succinylbenzoate, **5.38**, and thence into 1, 4-di-hydroxynaphthoate, **5.39**, prior to reaction with a variety of isoprenoid pyrophosphate units (Fig. 5.13).

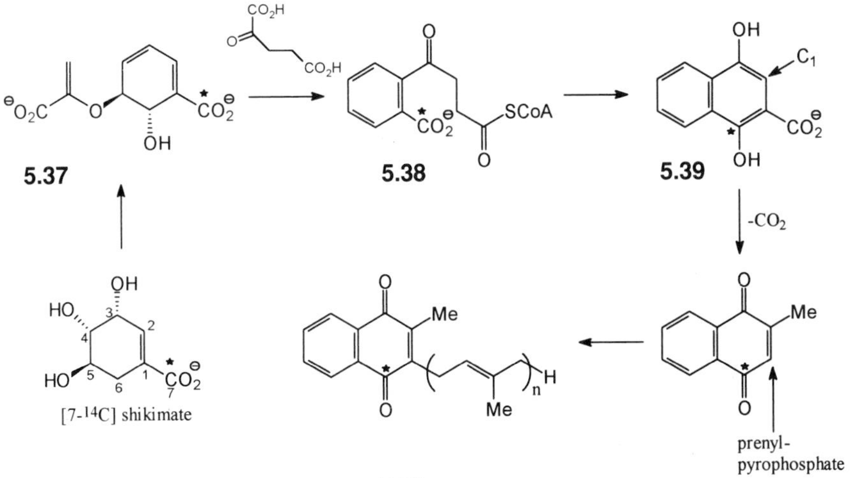

Fig. 5.13

6 Alkaloids

6.1

6.2

6.3

6.4

6.5

The alkaloids are, by definition, non-peptidic and non-nucleosidic compounds containing nitrogen. They are particularly abundant in higher plants, insects, amphibians, and fungi, and much less common in mammals. Typical examples include cocaine, **6.1**; nicotine, **6.2**; mescaline, **6.3**, an hallucinogenic constituent of the peyote cactus; morphine, **6.4**; and strychnine, **6.5**. Most alkaloids have biological activity of some kind, and many have been exploited for their pharmacological properties. But in the wild they probably act as feeding deterrents or mediators of other types of chemical communication between organisms, that is they are mediators of ecological interactions.

They are mostly produced from amino acids or closely related compounds, and it is assumed that they first evolved at a time when the organisms had a surplus of amino acids for their primary metabolic requirements. The new enzymes that were produced were involved in novel biosynthetic activities that initially served as routes for the destruction of the surplus amino acids. Those pathways that produced alkaloids with deterrent activity would help to ensure the survival of the organism, and were thus passed on (via genetic information) to the next generation.

Although some alkaloids are derived via insertion of ammonia (or some chemical equivalent of ammonia) into a preformed biosynthetic product, e.g. the biosynthesis of coccinelline **3.17** shown in Fig. 3.11, these compounds will not be considered. It is thus convenient to classify the majority of alkaloids according to their amino acid progenitors.

6.1 Alkaloids from ornithine and lysine

These two amino acids, ornithine, **6.6**, and lysine, **6.7**, are primarily involved in the biogenesis of aliphatic alkaloids, and the early stages of the biosynthetic pathways are probably similar (Fig. 6.1). The amino acids suffer decarboxylation to yield the diamines putrescine, **6.8**, and cadaverine, **6.9** (see Fig. 6.1) (the names reflect the 'aromas' of these compounds!), which then undergo methylation (S-adenosylmethionine) and 'transamination' (amine to aldehyde) before intramolecular condensation to form the imminium salts **6.10** and **6.11** (see Fig. 6.1). The mechanism of this transamination is shown in Fig. 6.2, and involves the cofactor pyridoxal phosphate in conjunction with various transaminase enzymes.

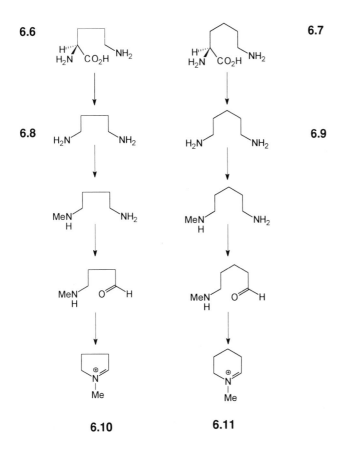

Fig. 6.1

The order of these reactions probably differs slightly between plant species, and the free (i.e. non-enzyme-bound) symmetrical diamines are not always true intermediates. However, for the biosynthesis of cocaine, **6.1**, in the South American plant *Erythroxylum coca*, the intermediacy of the symmetrical diamine was demonstrated by using [5-^{14}C]-ornithine as a precursor. The resultant cocaine (when degraded) was shown to have equal amounts of ^{14}C label at the positions C-1 and C-5. More recently, Leete has shown that the imminium salt, **6.10** (see Fig. 6.3) (labelled with ^{15}N, ^{13}C, and ^{14}C isotopes), was efficiently incorporated by the plant into cocaine. He went on to show that the remaining four carbons are derived from two successive additions of acetate (presumably in the form of acetyl-SCoA) to form 2-carbomethoxy-3-tropinone, **6.12** (Fig. 6.3), and that this proposed intermediate (labelled with ^{14}C) was also efficiently incorporated by the plant to produce cocaine. This efficiency was increased dramatically if the N-acetylcysteamine thioester of benzoic acid, **6.13** (a model for benzoyl-SCoA) was fed at the same time. So although the enzymology of the biosynthesis remains to be investigated, we can conclude that the proposed biogenesis shown in Fig. 6.3 is probably accurate.

PhCO.SCH$_2$CH$_2$NHAc

6.13

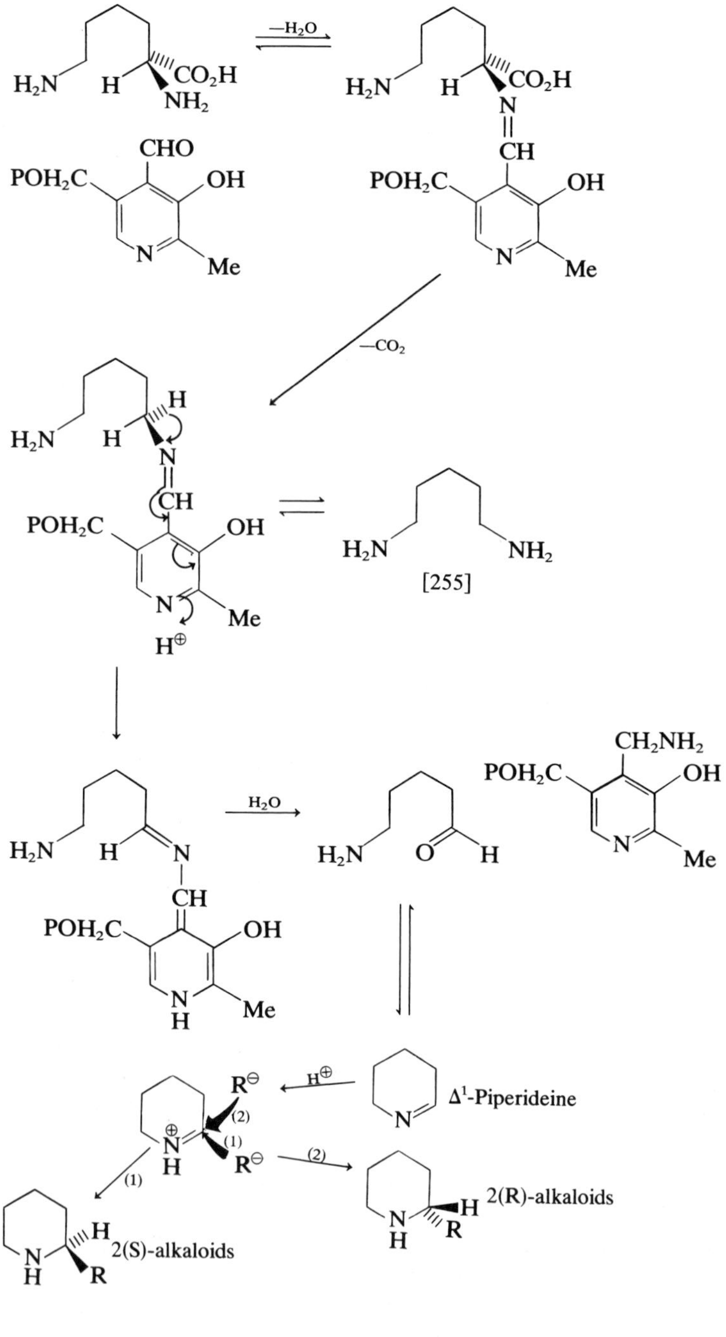

Fig. 6.2

6.10

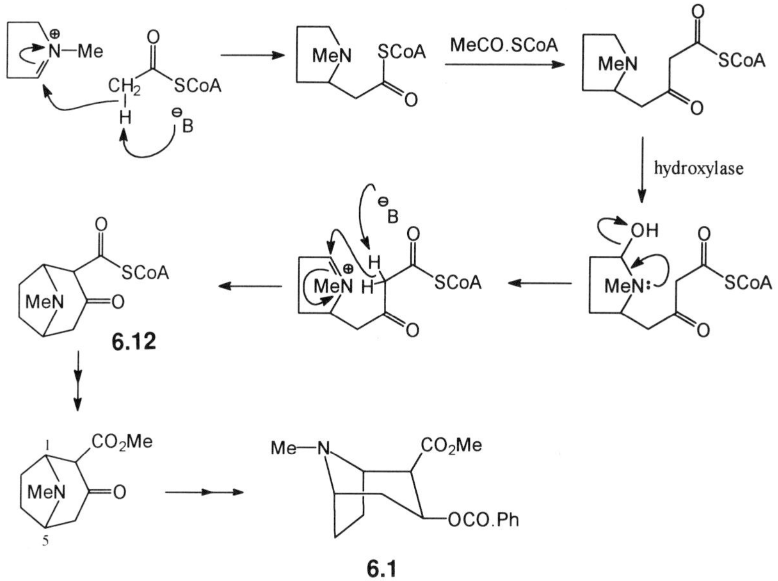

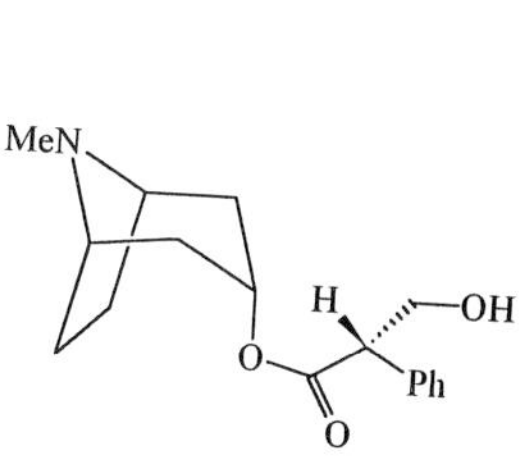

6.12

6.14

6.1

Fig. 6.3

The enzymology of the biosynthesis of hyoscyamine, **6.14**, has been studied in root cultures of *Atropa belladonna* (deadly nightshade). High levels of putrescine N-methyl transferase were found, but there was no evidence for the presence of ornithine-δ-N-methyl transferase or δ-N-methylornithine decarboxylase. This supports the involvement of free putrescine and N-methylputrescine in the biosynthesis of hyoscyamine, at least in this plant. In the various *Datura* species that have been studied, hyoscyamine is derived via a pathway that involves a non-symmetrical diamine, since only one bridgehead carbon atom was labelled by incorporation of [2-^{14}C]-ornithine. However, recent studies by Spenser suggest that the extra carbon atoms are derived, as with cocaine, from two separate molecules of acetyl-SCoA. These slight inter-species variations in biosynthetic pathways are quite common in secondary metabolism, and reflect subtle changes in enzyme levels during the evolution of the separate plant species.

These tropane alkaloids have a long association with magic and witchcraft. Coca leaves have been used as a source of cocaine for at least 2000 years, though more as a general stimulant and appetite suppressant than a psychomimetic. An estimated 8 000 000 South Americans still chew quids made from the leaves on a daily basis. The Incas revered coca leaves and certainly used them in magico-religious ceremonies—to commune with their gods. The conquistadores introduced the use of coca extracts into Europe, and various alcoholic beverages spiked with coca became popular for medicinal and recreational uses. Cocaine appears to function by inhibiting the re-uptake of the neurotransmitter dopamine (*vide infra*), and excessive amounts of this excitatory amine are thus available.

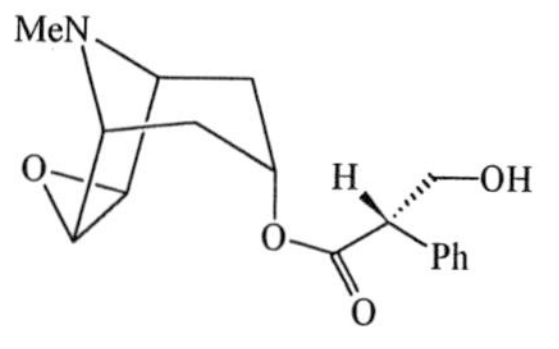

6.15

Hyoscyamine **6.14** and the more potent hyoscine (scopolamine), **6.15**, in contrast, bind to acetylcholine receptors (thus blocking access by acetylcholine), and have euphoriant and anaesthetic properties. In Ancient Greece extracts of the mandrake (*Mandragora officinarum*), a rich source of hyoscine, were valued for their anaesthetic properties, though larger doses were known to kill. The arch poisoners of mediaeval times certainly employed it for this purpose; but it was the witches and magicians who exploited its euphoriant properties to greatest effect. Before departure for the Sabbat, they 'smeared themselves under the arms and in other hairy places', thus ensuring that the hyoscine penetrated into the bloodstream via the sweat glands and fine blood capillaries. This avoided the more dangerous and less efficient oral route. As a result quite high concentrations of hyoscine built up in the brain, and at this level disorientation and hallucination would have occurred. The witches thus really believed that they 'flew' (on their broomsticks). These effects result from a blockade of the muscarinic class of receptors for the neurotransmitter acetylcholine, and this also causes a reduction in the flow of various bodily secretions. This is the main contemporary use for hyoscine, in premedication prior to surgery.

Hyoscine is also used in the prevention of sea- and airsickness in the form of alkaloid-impregnated plasters that allow percutaneous absorbtion of the drug, a variation of the route employed by witches. Due to these various therapeutic uses, the commercial production of hyoscine is thus of some importance. This should be facilitated by some recent genetic engineering, whereby extra copies of the genes coding for the enzyme hyoscine 6-hydroxylase have been incorporated into the genome of *A. belladonna*. This results in plants that have leaves and stems containing hyoscine as their (almost) exclusive tropane alkaloid, thus allowing direct extraction and purification by crystallization rather than by chromatographic separation.

The biosynthesis of the piperidine alkaloids, for example N-methylpelletierine, **6.16** (see Fig. 6.4), from lysine is broadly similar. The recent labelling experiment shown in Fig. 6.4 provides good evidence for the derivation of the 3-carbon unit from acetoacetyl-SCoA rather than from two moles of acetate, as with the tropane alkaloids.

The pyrrolizidine alkaloids are also derived from ornithine, and this interesting class of natural products contains compounds like retronecine, **6.17**, and the derived diester senecionine, **6.18** (see Fig. 6.5). Many of these are potent hepatotoxins (i.e. liver toxins), and there are frequent instances of livestock poisoning after ingestion of, for example, ragwort. The culprits appear to be pyrrole esters of general structure **6.19**, which are metabolites

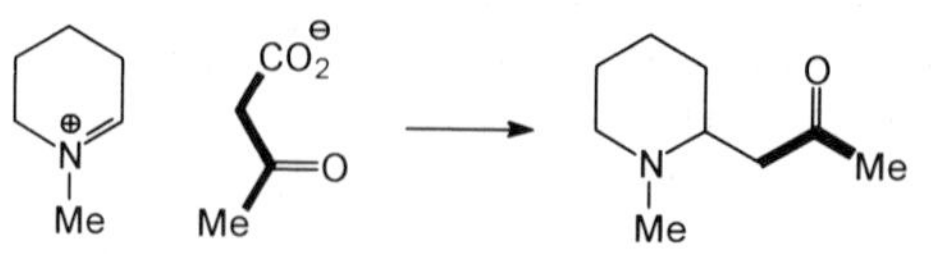

[1,2,3,4-^{13}C$_4$]-acetoacetate

Fig. 6.4

6.16

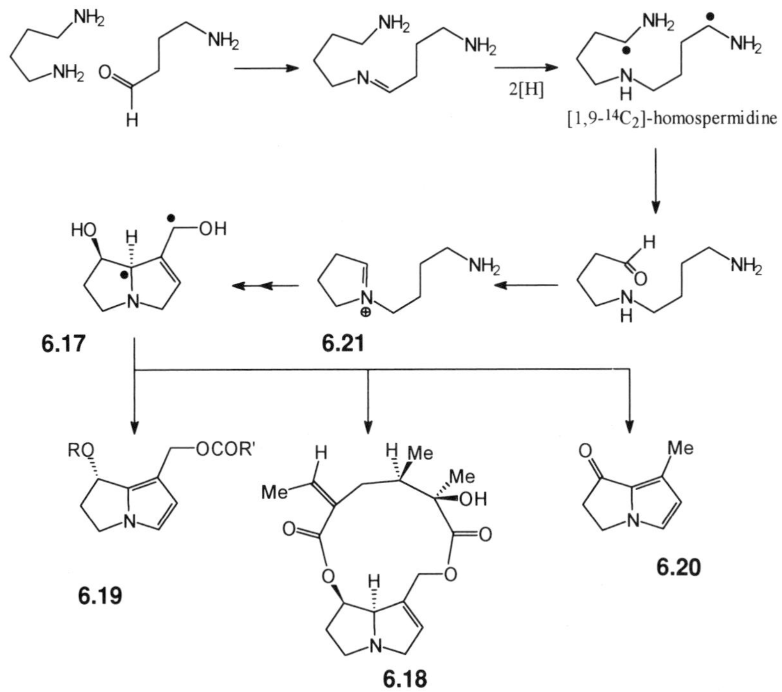

Fig. 6.5

of the pyrrolizidine alkaloids. These are potent alkylating agents and presumably damage enzymes and/or nucleic acids. Certain insects can store pyrrolizidine alkaloids as a means of making themselves distasteful to predators, and others use them as starting materials for the biosynthesis of attractants and aphrodisiacs. For example the modified alkaloid **6.20** is used by males of the Danaid family as a 'flight arrestant' for the female, as a prelude to mating.

The biosynthesis of these alkaloids has been exhaustively studied by Robins, and proceeds via the pathway shown in Fig. 6.5. The intermediacy of the pyrrolidinium ion **6.21** has been convincingly demonstrated by its incorporation (in radiolabelled form) into a variety of pyrrolizidine alkaloids.

Finally, in this section on alkaloids derived from aliphatic amino acids, brief mention should be made of the tobacco alkaloids, notably nicotine, **6.2**. This is derived from ornithine via the N-methylpyrrylium ion **6.10** in combination with nicotinic acid, **6.22** (Fig. 6.6). The comprehensive biosynthetic investigations by Leete and co-workers have provided evidence for the pathway shown in Fig. 6.6. A hydride is supplied by NADPH to nicotinic acid, and at the end of the pathway stereospecific loss of hydride (tritiide in the figure) and transfer to NADP$^+$ occurs. Nicotine has potent neuroactivity, primarily because it can interact with one class of acetylcholine receptors—so-called nicotinic subclass. As well as its obvious (addictive) properties in cigarette smoke, it has useful insecticidal activity as a neurotoxin, and tobacco waste is converted into nicotinium salts, **6.23**, for use as a commercial insecticide.

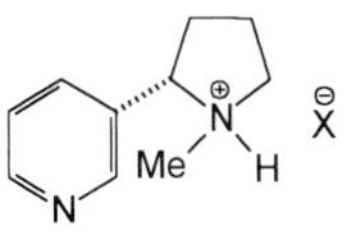

6.23

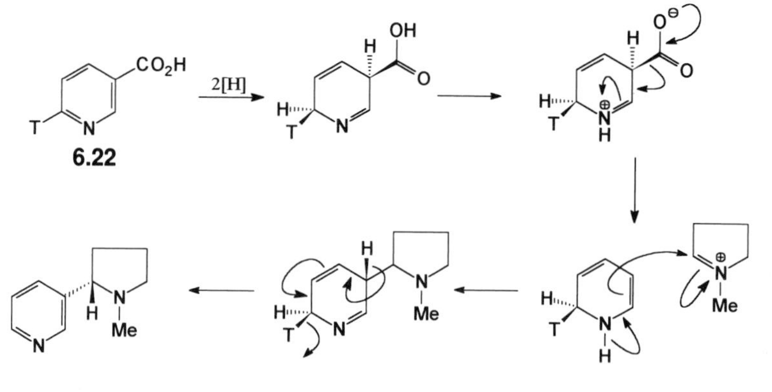

Fig. 6.6

6.2   Alkaloids from phenylalanine and tyrosine

This group of alkaloids has structures comprising ArC_2N units derived from tyrosine (**5.3**) together with, in most instances, ArC_2 or ArC_1 subunits derived from partial degradation of this amino acid. Representative examples included simple structures like mescaline, **6.3**, and the isoquinoline lophocereine, **6.24** (see Fig. 6.7), both produced by the peyote cactus; the benzylisoquinolines morphine, **6.4**, from the opium poppy and berberine, **6.25**, (see Fig. 6.13) from various *Berberis* species; and the Amaryllidaceae alkaloid lycorine, **6.26** (see Fig. 6.12) from plants of the daffodil family. The early stages of the biosynthesis of all of these alkaloids are similar, and involve decarboxylation of tyrosine to produce tyramine, **6.27**, which is then modified by further hydroxylation and/or methylation to produce the final metabolite (e.g. the pathway to mescaline shown in Fig. 6.7). Alternatively, the various arylethylamines may condense with an aldehyde (probably 3-methylbutanal in the biosynthesis of lophocereine, and certainly 4-hydroxyphenylethanal in the biosynthesis of (*S*)-reticuline, **6.28**) to produce the isoquinoline and benzylisoquinolines (Fig. 6.7). All of the enzymes on the pathway to reticuline have now been isolated and characterized, but it is interesting to recall that a biomimetric synthesis (Fig. 6.8) of the alkaloid norlaudanosine, **6.29**, was accomplished in 1930, thus establishing the feasibility of the chemistry before the biosynthesis had been investigated. It should also be added that biogenetic speculation in the last century and the early part of this century was often a key theoretical part of structure elucidation.

Another simple alkaloid with interesting pharmacology is ephedrine, **6.31** (see Fig. 6.9). This is produced by a number of *Ephedra* species, and the ancient Shennung herbal (100–200 AD, though many of the entries can be dated to a much earlier epoque) recorded the use of a preparation known as 'Ma Huang' for the treatment of respiratory conditions like bronchitis and asthma. The structure of ephedrine was elucidated in 1923, and it entered

The peyote cactus, *Lophophora williamsii*, is a source of a number of potent hallucinogens, e.g. mescaline, and forms the basis of the magical preparation known as 'peyotl'. This has been used by the Indians of the southern part of the USA, Mexico, and Central America for magico-religious purposes for several thousand years. The peyote cult is still widely practised and was legalized by the US Congress in 1967. A graphic account of the psychomimetic effects of mescaline were given by Aldous Huxley in his book *The Doors of Perception*, in which he claimed, somewhat fancifully, that the drug allowed the user to 'experience only the heavenly side of schizophrenia'. Like most hallucinogenic substances, mescaline probably exerts its effects by binding to brain receptors that would otherwise bind natural neurotransmitters like dopamine (dihydroxyphenylethylamine, **6.30** (Fig. 6.7).

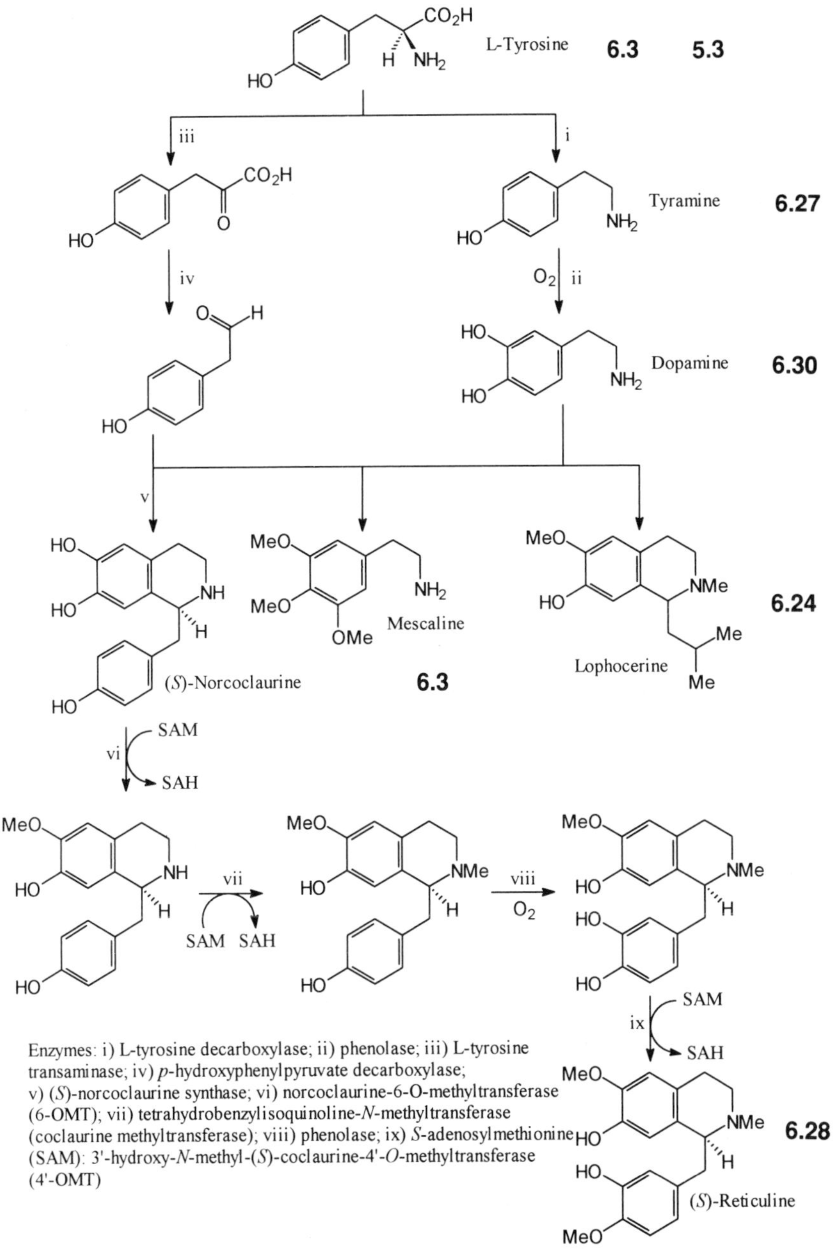

Enzymes: i) L-tyrosine decarboxylase; ii) phenolase; iii) L-tyrosine transaminase; iv) *p*-hydroxyphenylpyruvate decarboxylase; v) (*S*)-norcoclaurine synthase; vi) norcoclaurine-6-O-methyltransferase (6-OMT); vii) tetrahydrobenzylisoquinoline-*N*-methyltransferase (coclaurine methyltransferase); viii) phenolase; ix) *S*-adenosylmethionine (SAM): 3'-hydroxy-*N*-methyl-(*S*)-coclaurine-4'-*O*-methyltransferase (4'-OMT)

Fig. 6.7

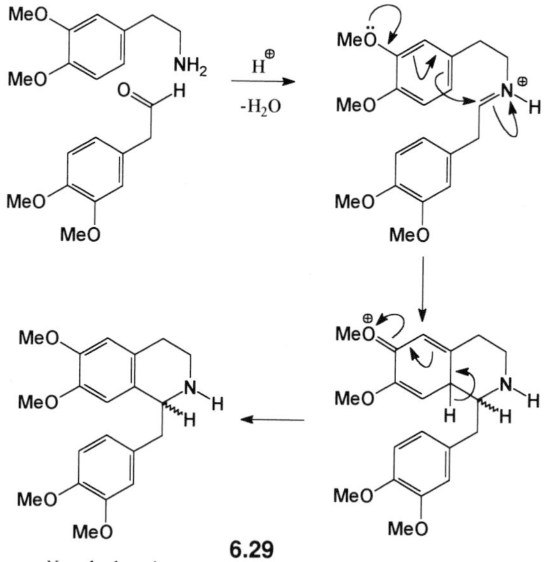

6.29

N-norlaudanosine

Fig. 6.8

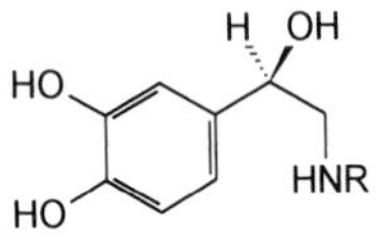

clinical use in 1926 as a bronchodilator for the treatment of asthma. Unfortunately it also stimulates the heart, which is not surprising given its structural similarity to the 'fight or flight' hormone adrenaline, **6.32** and the neurotransmitter noradrenaline, **6.33** (both produced by the adrenal glands). The pharmaceutical industry subsequently spent enormous efforts on the design of bronchodilators devoid of the cardiac stimulant effect. The highly successful antiasthma drugs salbutamol (Ventolin), **6.34**, and terbutaline (Bricanyl), **6.35**, were the result of this research.

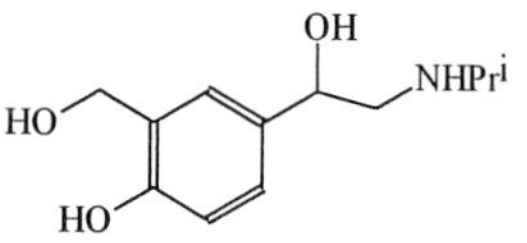

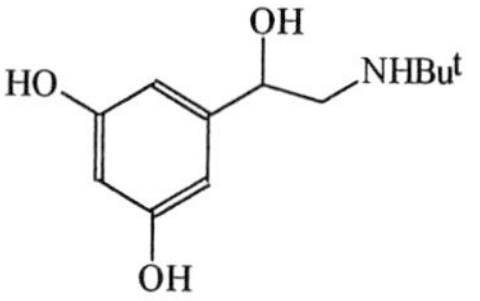

The biosynthesis of ephedrine is not as simple as it appears, and it has recently been shown that the structure is probably produced from benzoic acid and pyruvic acid (Fig. 6.9). Use of $[2, 3-^{13}C]$-pyruvate led to intact incorporation of the C-2 to C-3 moiety, and the proposed biogenesis is shown in the figure.

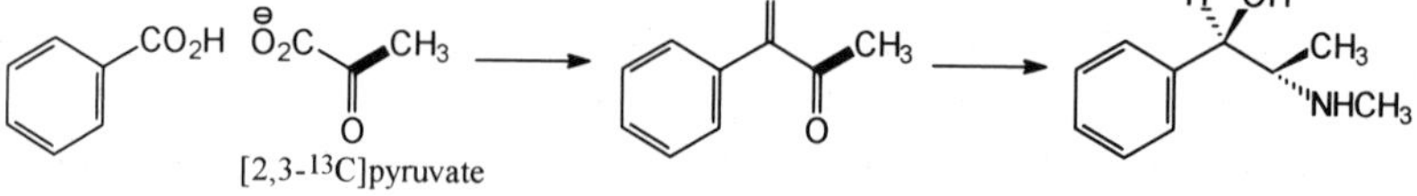

6.32,   R = Me
6.33,   R = H

6.34

6.35

$[2,3-^{13}C]$pyruvate

6.31

Fig. 6.9

The opium alkaloids also have a long history of human use. Crude opium is, by definition, the air-dried, milky exudate obtained from unripe seed capsules of the poppy *Papaver somniferum*, and comprises about 25 per cent by weight of alkaloids. The biosynthesis of the opium alkaloids has been studied with isotopically labelled substrates (Fig. 6.10), and more recently

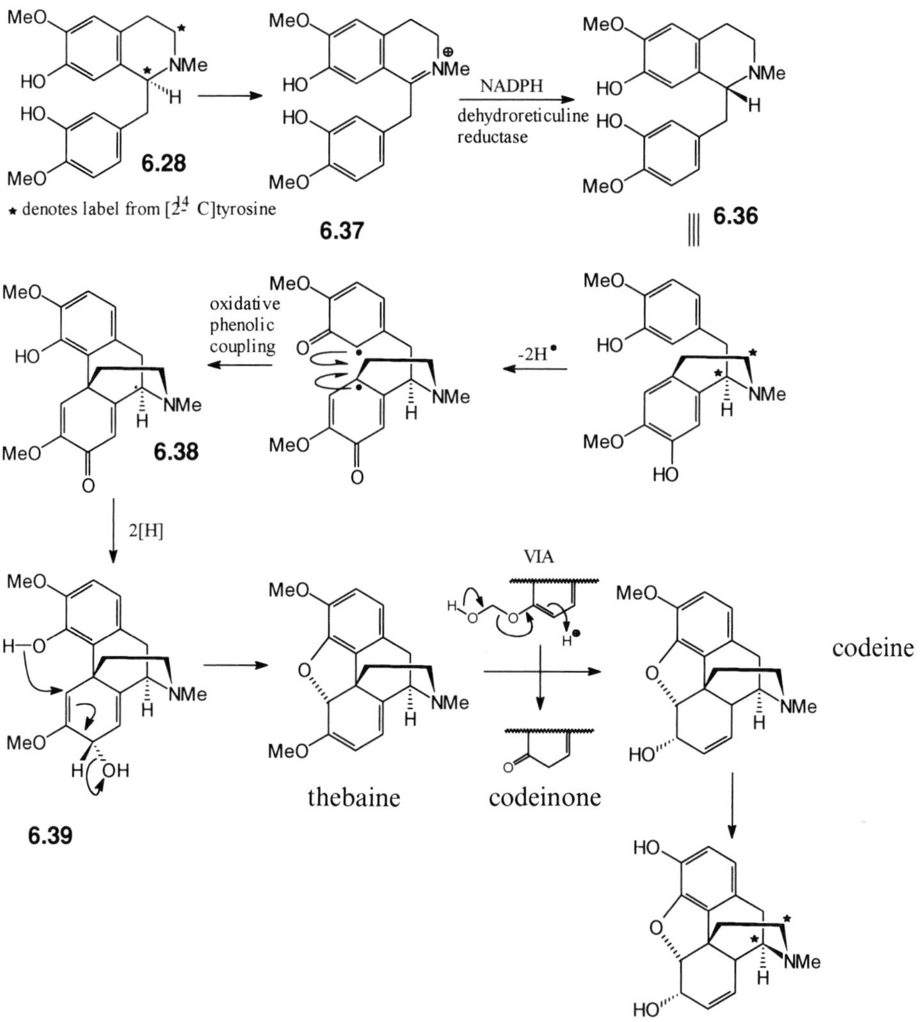

* denotes label from [2-^{14}C]tyrosine

Fig. 6.10

by characterization of the discrete enzymes involved. Intriguingly the (*S*)-reticuline (**6.28**) must be epimerized to (*R*)-reticuline, **6.36**, via the immonium ion **6.37** prior to its conversion into salutaridine, **6.38**, by means of oxidative phenolic coupling. The cytochrome P_{450} enzyme that catalyses this process has been isolated recently. After reduction to salutaridinol, **6.39**, the oxygen bridge is formed via an S_N2' reaction to produce thebaine and the proposed chemistry for the demethylation of the methoxyl group to produce codeinene is also of interest. Further functional group alterations provide codeine, and thence morphine.

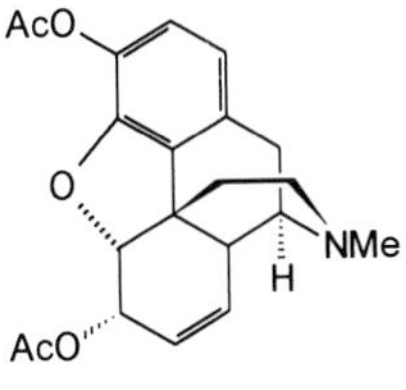

The analgetic (painkilling) and addictive properties of morphine and the synthetic derivative heroin, **6.40**, are well known. In the early 1970s, labelling studies with tritiated morphine demonstrated the existence of specific binding sites for morphine in the brain; and at least three different classes of opiate receptor have now been identified. This immediately poses the question: why does the brain possess receptors for a plant alkaloid? The obvious explanation is that the brain produces its own endogenous 'opiate-like' compounds, and the existence of these was first demonstrated in 1975. Two pentapeptides termed enkephalins were first isolated, and later the parent

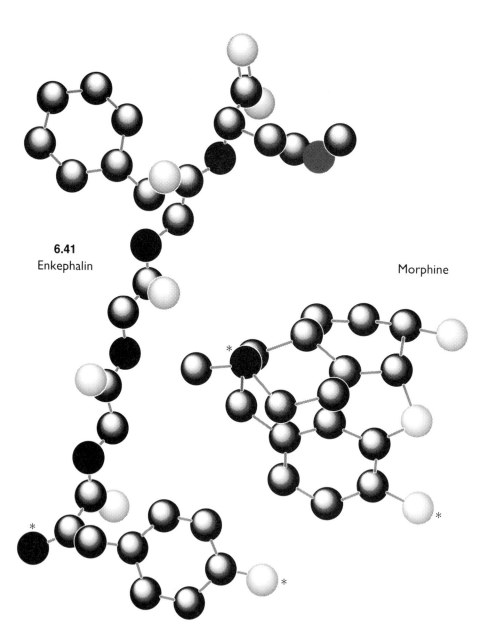

6.41
Enkephalin

Morphine

Fig. 6.11   The similar structural relationship between the marked (∗) atoms in the two molecules probably accounts for their similar pharmacological properties.

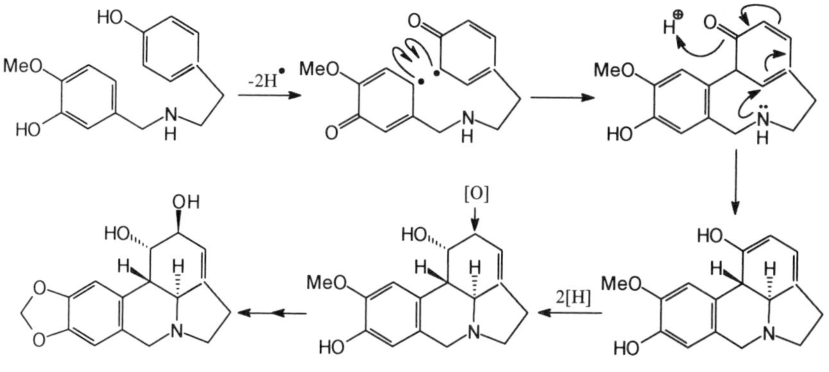

6.26   **6.42**

Fig. 6.12

polypeptides (endorphins) were also identified. These bind to the opiate receptors and appear to control levels of pain in the body, as well as, not surprisingly, having an effect on mood, eating habits, etc. The structural similarity between morphine and the N-terminus of methionine enkephalin, **6.41**, is shown in Fig. 6.11. An oxidative phenolic coupling process is also implicated in the biosynthesis of other alkaloids, as for norpluviine, **6.42**, and lycorine, **6.26** (Fig. 6.12).

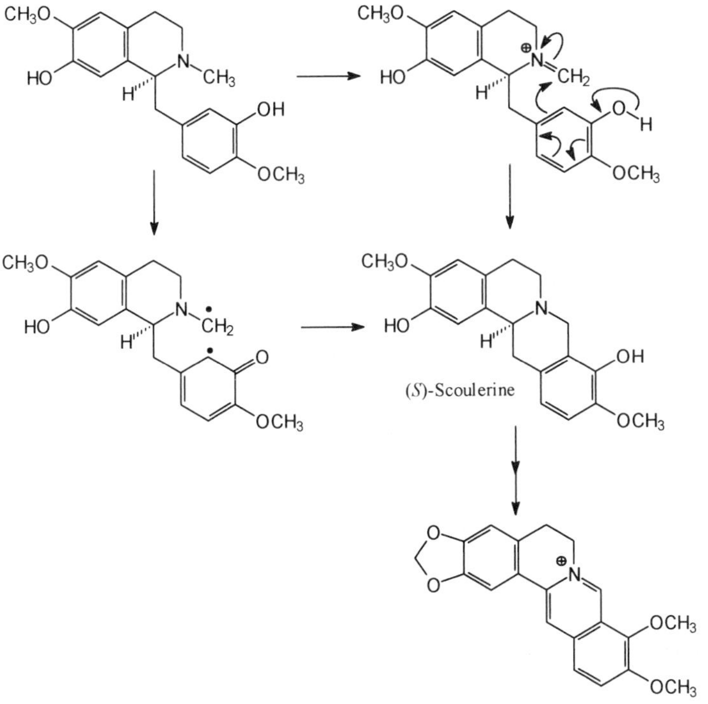

(*S*)-Scoulerine

6.25

Fig. 6.13

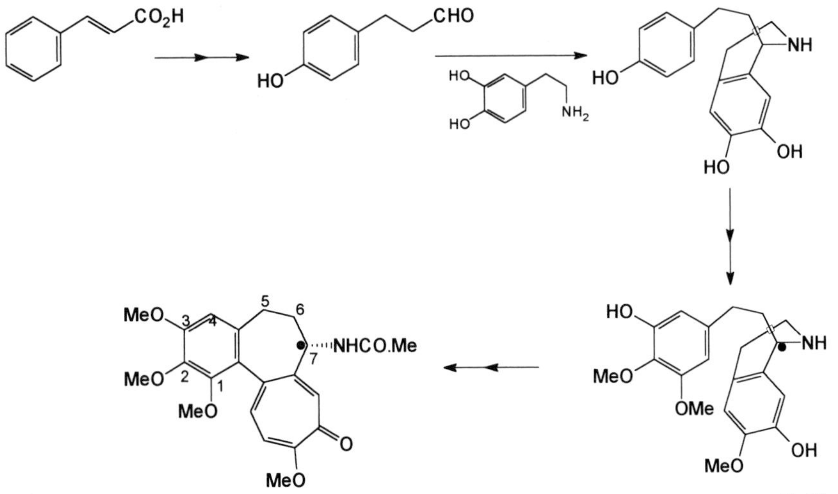

Fig. 6.14   **6.43**   **6.44**

Colchicine probably alleviates the symptoms of gout by inhibiting various enzymes involved in the inflammatory process, but it also has cytotoxic properties, and has been considered for use as an anticancer agent. It binds to and inhibits the enzyme tubulin polymerase which catalyses the formation of the protein tubulin. This protein provides the structural support (the mitotic spindle) that holds the two developing cells together during cell division (mitosis). Colchicine itself is probably too toxic for clinical use, but it has proved very useful as a tool for studying the process of mitosis.

Further extensive studies (by Zenk and co-workers) of the enzymes of benzylisoquinoline biosynthesis have demonstrated the discrete existence of the various hydroxylases and N- and O-methyltransferases involved in the biosynthesis of berberine, **6.25**. Two possible mechanisms for the formation of the so-called 'berberine bridge' are shown in Fig. 6.13.

Finally in this section, brief mention should be made of the alkaloid colchicine, **6.43**. Extracts of the autumn crocus, *Colchicum autumnale*, were used by the ancient Romans as a treatment for gout. This is now known to be caused by a disturbance in metabolism of nucleic acids, with resultant deposition of crystals of uric acid in the joints.

The biosynthesis of colchicine has been extensively studied, and probably proceeds via the pathway shown in Fig. 6.14. The mechanism of the later stages of the biosynthesis are still speculative, but most of the early intermediates have been isolated. $[1-^{13}C]$-autumnaline, **6.44**, has been incorporated into colchicine, and the enhancement of the signal due to C-7 of colchicine (Fig. 6.14) provides cogent evidence for the pivotal role of this intermediate.

6.3 Alkaloids derived from tryptophan

The amino acid tryptophan, **5.4**, and its decarboxylation product tryptamine, **6.46** (see Fig. 6.15), are the precursors of literally thousands of alkaloids. Their skeleta are easily discernible within the structures of the hallucinogen pscilocin, **6.47**, from the *Pscilocybe* mushroom and physostigmine, **6.48**, from the Calabar bean; rather less obviously within the structures of lysergic acid, **6.49** (Fig. 6.16) and strychnine, **6.5**. The probable biogenesis of physostigmine is shown in Fig. 6.15 and the mechanistic feasibility of this is supported by the biomimetic synthesis of deoxypseudophyrynaminol, from the skin of an Australian frog, shown in Fig. 6.17.

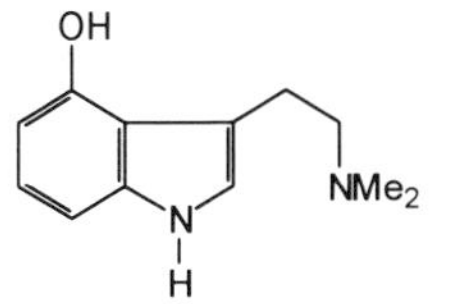

6.47

Extracts of the Calabar bean were once the basis for trials by ordeal, whereby a supposed criminal was forced to consume the extract to prove innocence or guilt. Rapid consumption of the brew probably resulted in vomiting, thus denying access to the bloodstream for the neurotoxic physostigmine. In this way the truly innocent may have survived the ordeal. The alkaloid is a potent inhibitor of the enzyme acetylcholine esterase, which hydrolyses the neurotransmitter acetylcholine once a signal has been conveyed. Clearly inhibition of this enzyme will lead to a rise in concentration of acetylcholine and resultant neurotoxicity. Though it has proven clinical utility in myasthemia gravis, a condition in which the number of acetylcholine receptors are much reduced. And it is under evaluation for the treatment of Alzheimer's disease in which there are reduced levels of acetylcholine in the brain.

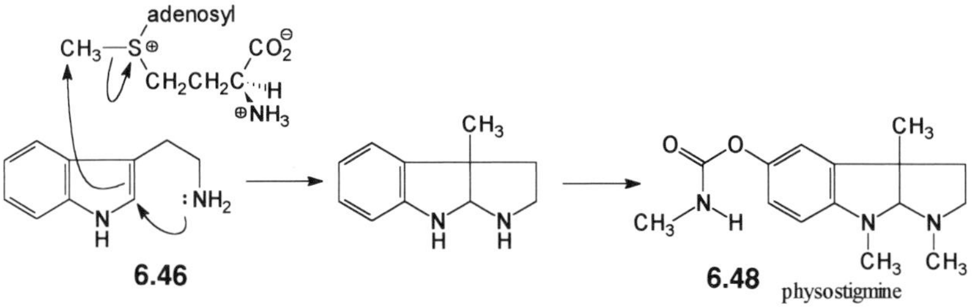

Fig. 6.15

The ergot alkaloids, of which lysergic acid, **6.49**, is the parent structure, are produced via the pathway shown in Fig. 6.16. The intricate details of some of the stages are not yet known, but elegant isotopic studies have established the details shown. The initial step involves an entirely reasonable electrophilic aromatic substitution reaction between tryptophan and dimethylallyl pyrophosphate. The anticipated inversion of configuration has been confirmed through the use of specifically tritiated MVA. Hydroxylation of this intermediate, **6.50**, probably ensues, because this compound (in radiolabelled form) has been shown to be efficiently incorporated into elymoclavine, **6.51**, by cultures of *Claviceps purpurea*, a fungus that produces a large number of ergot alkaloids. The rest of the sequence to chanoclavine-I, **6.52**, is speculative, though the rotation about the terminal isopropyl group has been established using [13]-C-labelled dimethylallyl pyrophosphate. Specific oxidation of one of the methyl groups produces **6.52**, and further oxidation to the aldehyde, condensation with the amine and hydroxylation produces elymoclavine, **6.51**, and thence lysergic acid.

The fungus *Claviceps purpurea* is a common contaminant of cereals, especially rye and barley, and in the Middle Ages consumption of contaminated bread was responsible for the numerous plagues of 'ergotism' that were recorded. Several of the ergot alkaloids, most notably ergometrine,

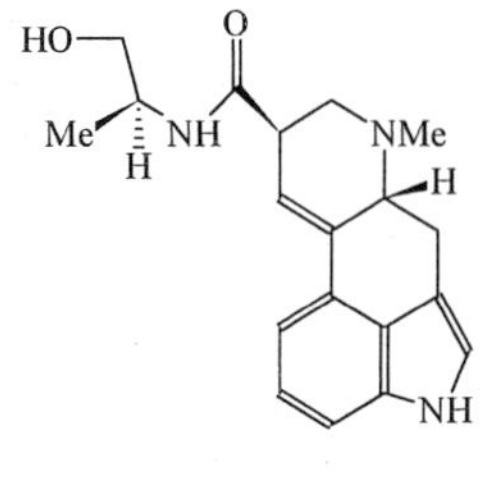

6.53

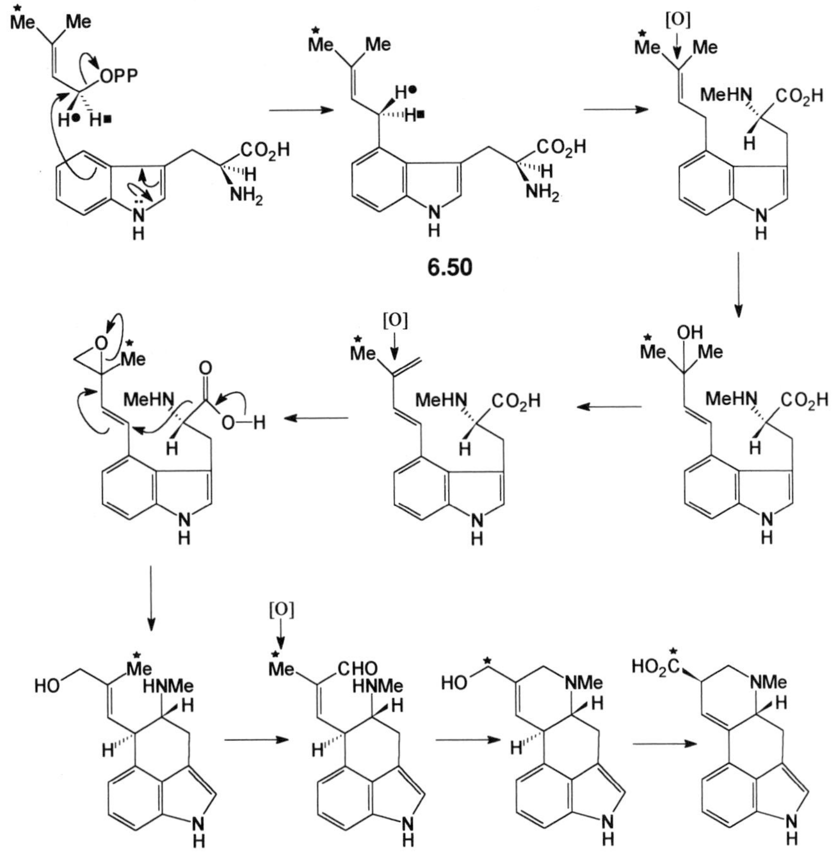

6.52 6.51 6.49

Fig. 6.16

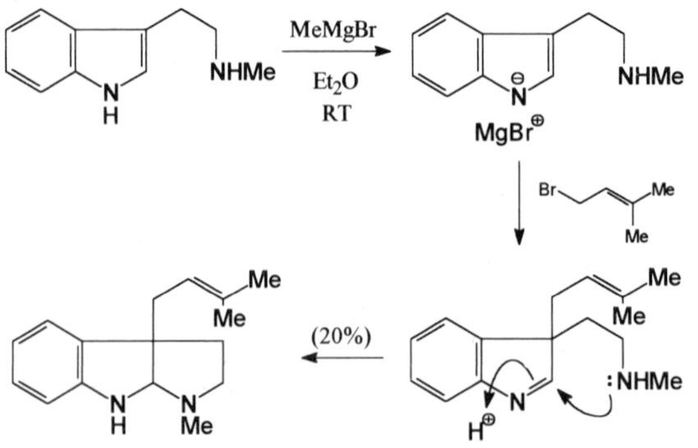

deoxypseudophrynaminol

Fig. 6.17

6.53, have potent vasoconstrictive activity, and induced gangrene of the extremities in those affected. The blackened limbs were believed to have been consumed by 'holy fire', and the Church was often able to alleviate some of the misery of the condition, through the administration of uncontaminated bread, though they would have argued that the power of prayer was the ultimate saviour.

The ergot alkaloids attained a more recent notoriety following Albert Hofmann's discovery that the wholly synthetic lysergic acid diethylamide (LSD), **6.54**, had hallucinogenic activity in minute doses. He subsequently showed that the Aztec magical preparation ololuiqui, based upon an extract of the plant *Rivea corymbosa*, also contained simple lysergic acid amides, and thus explained an ethnopharmacological mystery that had existed since the time of the conquistadores.

The central nervous system activity of LSD and ololuiqui (and also psylocin, **6.47**) should come as no surprise given their structural similarity to 5-hydroxytryptamine (5-HT), **6.55**. This is a natural neurotransmitter, and is implicated in the control of sleep states and mood, and disturbances in the levels of 5-HT are known to be implicated in such conditions as migraine, schizophrenia, and dementia. Interestingly, one class of 5-HT receptor is

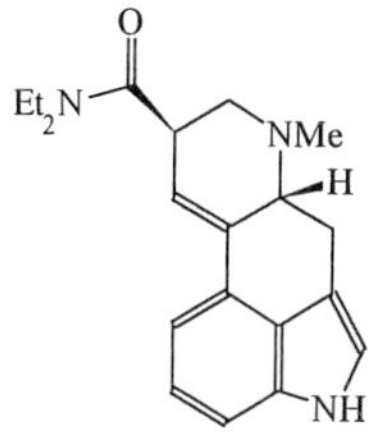

6.54

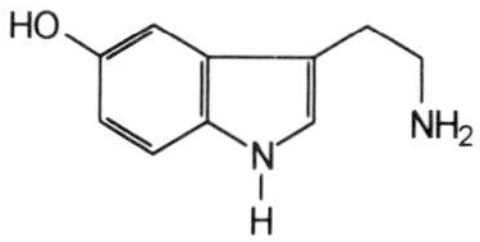

6.55

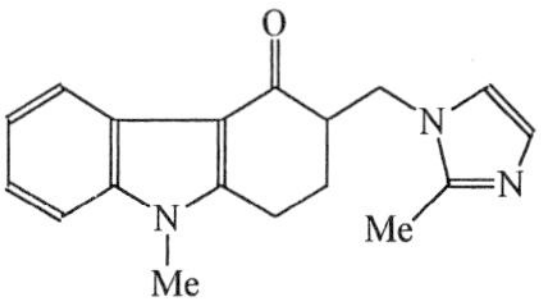

6.56

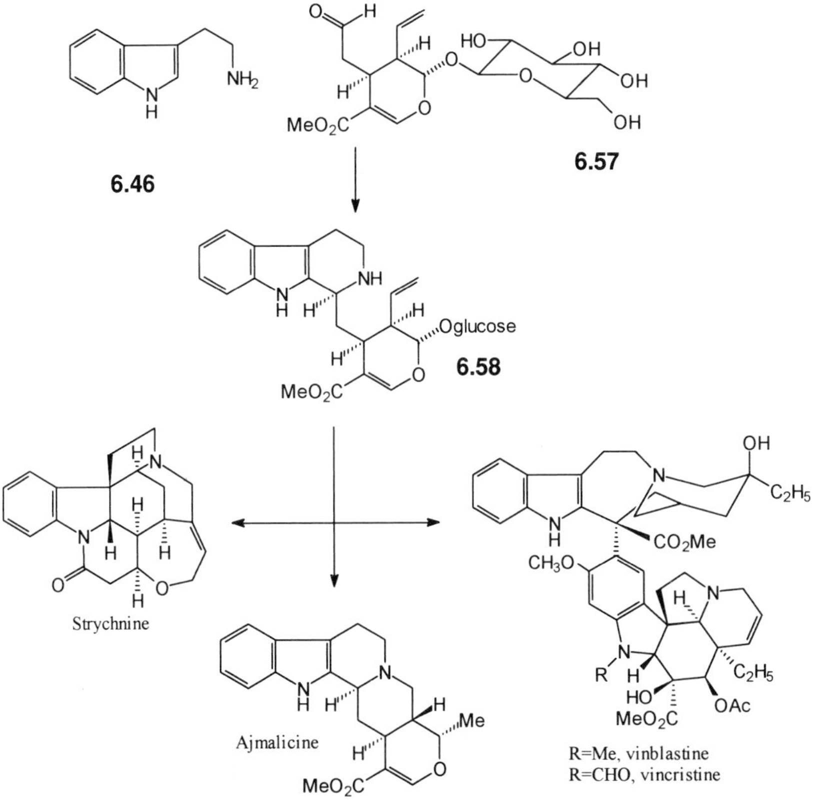

6.46

6.57

6.58

Strychnine

Ajmalicine

R=Me, vinblastine
R=CHO, vincristine

Fig. 6.18

also involved peripherally in the control of emesis (vomiting) caused by, for example, anticancer drugs. The new and highly effective antiemetic agent Ondansetron, **6.56**, has high affinity for these 5-HT$_3$ receptors.

The great majority of alkaloids containing an indole moiety arise via a pathway that involves condensation of tyramine, **6.46** and the highly oxygenated terpenoid secologanin, **6.57** (see Fig. 6.18). The product, 3 (*S*)-strictosidine, **6.58**, is the precursor of over 1200 different so-called 'monoterpene indole alkaloids', and some of these are depicted in Fig. 6.18. Details of the biosyntheses are beyond the scope of this book, but it is worth noting that work in underway on the enzymology and molecular biology of these pathways. For example, the enzyme strictosidine synthase from *Rauwolfia serpentina* has been isolated and fully characterized, and has also been immobilized on an inert support where it remains stable for long periods (half-life around 100 days). Work of this kind should soon allow the synthesis (in culture) of large quantities of specific alkaloids. This has obvious potential when one considers the enormous value of the broad-spectrum anticancer agents vinblastine and vincristine, or the potential antianxiety activity of ajmalicine.

6.4   Other amino acid metabolites

Many other amino acids are found as constituents of secondary metabolites, and two examples are shown opposite. Indolactam V, **6.59**, the precursor of the teleocidins, tumour-promoting agents from *Streptoverticillium blastmyceticum*, clearly incorporates fragments from tryptophan and valine. Preechinuline, **6.60**, a mycotoxin from *Aspergillus amstelodami* is produced from tryptophan, alanine, and dimethylallyl pyrophosphate. But the most studied amino acid metabolites are the β-lactam antibiotics, especially the penicillins. Although Baldwin and his respective co-workers continue to unravel the intimate details of their biosynthesis, a discussion of the various chemical speculations is beyond the scope of this book. However, there is no question that the overall pathway shown in Fig. 6.19 is accurate. One enzyme, isopenicillin-N-synthase, catalyses the conversion of the tripeptide δ-(L-alpha-aminoadipoyl)-L-cysteinyl-D-valine (hereafter LLD-ACV), **6.61**, into isopenicillin-N, **6.62**. The enzyme is known to require ferrous iron as a cofactor, and is not dependent upon the cofactor haem (as are the cytochrome P_{450} enzymes). Isotopic labelling studies have conclusively demonstrated that only four hydrogens are lost during the formation of isopenicillin-N (shown in the figure), and experiments with substrate analogues have revealed much about the structural and conformational requirements imposed by the enzyme. Thus the results shown in Fig. 6.20 demonstrate that substrate analogue **6.63** is an acceptable substrate, in that it leads to a penicillin analogue, whilst analogue **6.64** becomes involved in a 2π + 2π cycloaddition reaction and an ene-type reaction respectively to yield products **6.65** and **6.66**. Other biosynthetic investigations with isotopi-

6.59

6.60

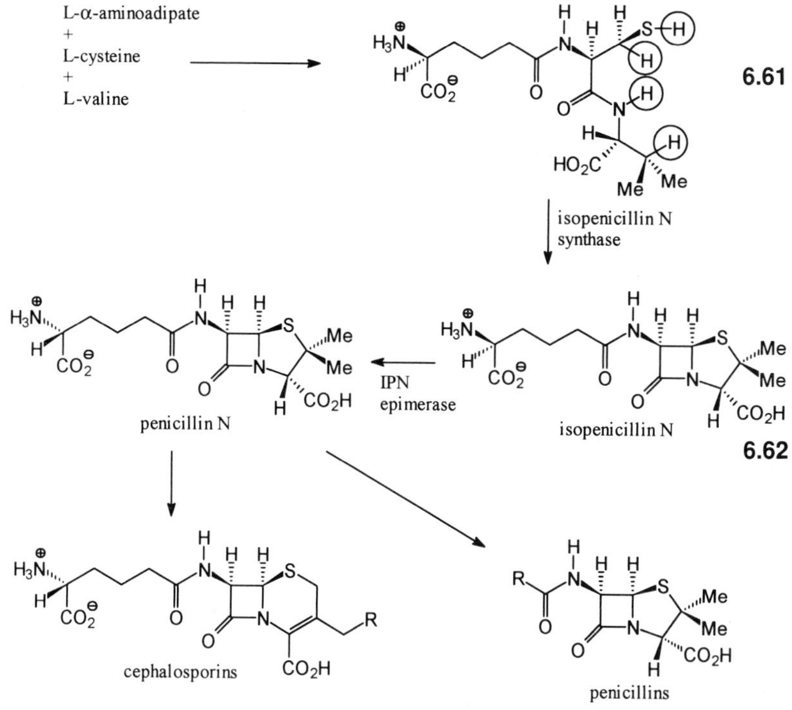

Fig. 6.19

cally labelled substrates and substrate analogues have been used to probe the mechanism of ring expansion implicated in the biosynthesis of the cephalosporin nucleus, and the interested reader is urged to consult the numerous papers by Baldwin for further details of these fascinating studies.

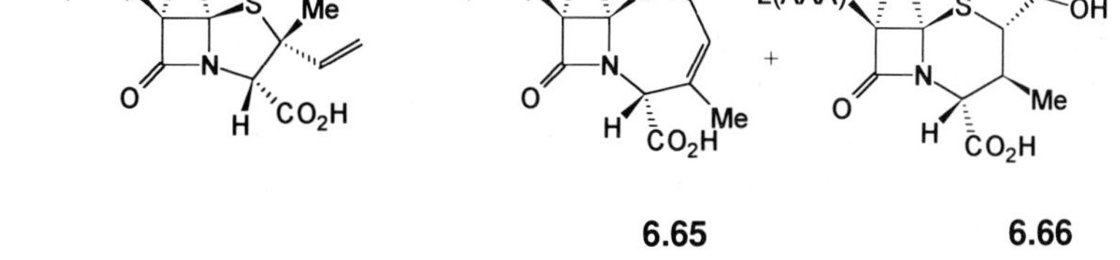

Fig. 6.20 L(AAA) ≡ L-alpha-aminoadipoyl.

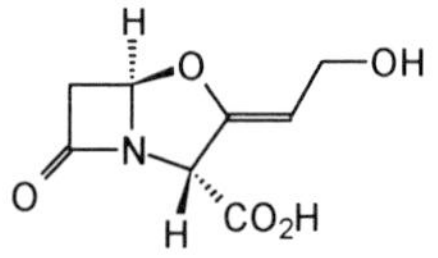

6.67

The story of Fleming's initial discovery of penicillins, and their development by the Florey and Chain group, is well known (see reading list for sources). They act by disrupting the assembly of new cell wall material by growing bacteria, with the β-lactam ring acting as a point of attack by the bacterial enzyme transpeptidase (Fig. 6.21). The enzyme is thus inhibited through formation of a covalent bond to the penicillin (or cephalosporin). Over the years the bacteria have evolved to produce new enzymes that hydrolyse the β-lactam ring by essentially the same mechanism. Other secondary metabolites, like clavulanic acid, **6.67**, have been discovered, which inhibit these β-lactamases, thus allowing the pharmaceutical industry to stay one step ahead of the bacteria.

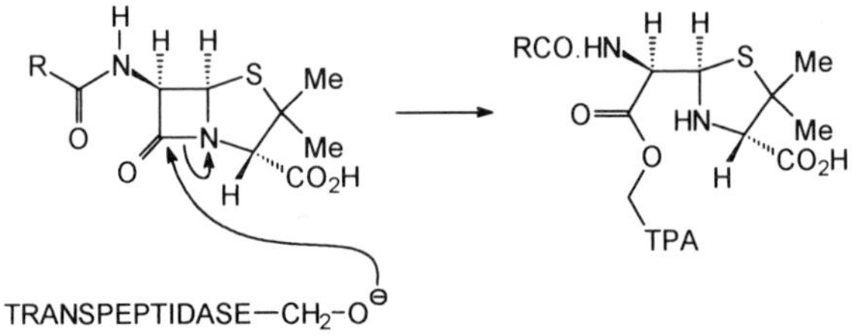

Fig. 6.21

7 Ecological chemistry

The use of secondary metabolites as mediators of interactions between organisms has been mentioned at several points in the previous chapters. During the past 20 years an increasing number of such inter- and intra-species interactions have been indentified, and this area of study is usually known as 'ecological chemistry' since the secondary metabolites help to establish and maintain the place or ecological niche of a species.

Speculation about the origins of the main secondary metabolic pathways was included in Chapter 1, and there is little doubt that the plethora of secondary metabolites that are produced by plants and insects (in particular) have arisen in response to the continually changing interactions that occur between species. Plants and insects have been co-evolving for millions of years, and as a result plants have developed biosynthetic pathways that produce toxic or distasteful compounds, whilst insects have devised means of overcoming these defences and even ways of using these deterrent chemicals for their own purposes.

Subtle changes in the levels of particular enzymes may alter the structure or quantity of a particular secondary metabolite that is produced, and this may ultimately lead to an improvement in the survival prospects of the producer species. Any small advantage that a species derives from its use of secondary metabolites may improve its tenuous hold upon a particular ecological niche, and its chances of mating and passing on its (improved) genes to its offspring.

Many types of interactions mediated by secondary metabolites have been identified, and the main classes are described in the following sections.

7.1 Plant–insect interactions

Many plants produce secondary metabolites that are distasteful or actually toxic to insects (and other herbivores including mammals). Perhaps the most obvious examples of such compounds are the cyanogenic glycosides, e.g. dhurrin, **7.1**, which are broken down in the mouthparts of the herbivore to produce HCN. Most alkaloids are distasteful, and compounds like nicotine, **6.2**, and retronecine, **6.17**, deter a wide variety of herbivores. For example, wild tobacco (*Nicotiana sylvestris*) can increase its production of nicotine by 3–4 times when under attack, and this is sufficient to deter most

7.1

7.2

7.3

7.4

insect species. However, the tobacco hornworm (*Manduca sexta*) has 'learnt' how to avoid triggering this response. It feeds without damaging the major leaf veins of the plant, with the result that only a small increase in nicotine production ensues.

A much more structurally complex antifeedant molecule is azadirachtin, **7.2**, produced by the Indian neem tree (*Azadirachta indica*). For centuries various parts of the plant have been used to protect plants and clothes from the attentions of insects. The structure of the active compound was proved in 1987. It is probably produced by modification of the steroidal intermediate tirucallol, **7.3**, but little biosynthetic work has been carried out. Its great potency and selective activity against insects have led to its inclusion in the first commercial antifeedant called 'Margosan-O'.

Often the secondary metabolites reduce the growth or maturation of the insects that feed upon a plant. The subtle hormonal effects of juvabione (**4.26**) and the ecdysones, e.g. **4.49**, were mentioned in Chapter 4, and sedge (*Cyperus iria*) actually contains high concentrations (*c.* 150 microgram per gram wet weight of leaf) of the insect juvenile hormone JHIII, **7.4**—an amount that is around 150 times that found in a typical insect. It certainly causes sterility in grasshoppers that feed on *C. iria*.

The sesquiterpene lactone tenulin, **7.5**, from *Helenium amarum*, is a potent antifeedant but also disrupts growth and development of insect larvae. Like most compounds that possess conjugated unsaturated carbonyl systems it probably participates in Michael reactions with 'biological nucleophiles' on (insect) enzymes. Support for this hypothesis has been provided by experiments in which tenulin and cysteine were coadministered to insect larvae, and the toxicity of the sesquiterpene was apparently reduced.

Not all interactions are deterrent in nature since plants often rely upon insects for assistance with pollination. Floral fragrances, usually monoterpenes or low molecular weight aromatic compounds, are important insect attractants, and some of the interactions are highly novel. Thus the pollination of *Datura innoxia* by various species of hawkmoths relies upon the narcotic properties of the constituent tropane alkaloids, e.g. hyoscine, **6.15**. These narcotize the moths which appear to become 'addicted' to the alkaloids, and return again and again to obtain further 'fixes' and this ensures efficient pollination.

7.5

Finally some interactions are exceeding complex, and a good example is provided by recent work with corn, *Zea mays*. In response to attack by insects, corn produces (3*Z*)-hexen-1-yl-acetate, **7.6**, linalol, **7.7**, and indole, **7.8**. These are attractive to the parasitic wasp, *Cotesia marginiventris*, which then preys upon the insects that caused the initial damage—an apparent example of a failed defence system that is none the less eventually beneficial to the plant.

7.6

7.7

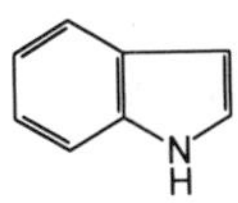

7.8

7.2 Insect–insect interactions

Most insects communicate with one another by means of pheromones. These are primarily of two types—sex pheromones which are usually released by the female as a lure for the male, and trail pheromones which are used by social insects (termites and ants) to mark trails from the nest to food sources. Some insects produce these compounds from simple precursors (acetate, mevalonate, etc.), whilst others require plant chemicals as starting materials for their biosynthetic activities.

Sex pheromones are almost invariably species-specific, and complex mixtures of secondary metabolites are often required to ensure that non-productive inter-species mating does not occur. Thus female cabbage looper moths (*Trichoplusia ni*) and soya bean looper moths (*T. includens*) use (7*Z*)-dodecenyl acetate as a major component of their sex pheromones. However, *T. ni* also releases three species-specific compounds—(5*Z*)-dodecenyl acetate, (7*Z*)-tetradecenyl acetate, and (9*Z*)-tetradecenyl acetate—and *T. includens* has two such compounds—(7*Z*)-dodecenyl propionate and (7*Z*)-dodecenyl butyrate. The bolas spider (*Mastophora cornigera*) produces a mixture of (9*Z*)-tetradec-9-en-1-yl-acetate, (9*Z*)-tetradec-9-enal, and (11*Z*)-hexadec-11-enal. These are female sex pheromones for a number of species of moth, and the spider is thus able to lure male moths to their deaths. There is even evidence that the spider can vary the composition of its 'attractive cocktail' to match the mating patterns of the various species during the year.

Male cotton boll weevils are attracted to cotton plants by a mixture of volatile monoterpenes released by the plants. They feed and then release an aggregation pheromone which comprises four ten-carbon metabolites (Fig. 7.1), and this causes a mass aggregation and subsequent mating and egg laying on the cotton plants. Presumably the monoterpenes released by the cotton plant were originally deterrent substances, but the boll weevil has evolved a means of using these as starting materials for its own biosynthetic activities.

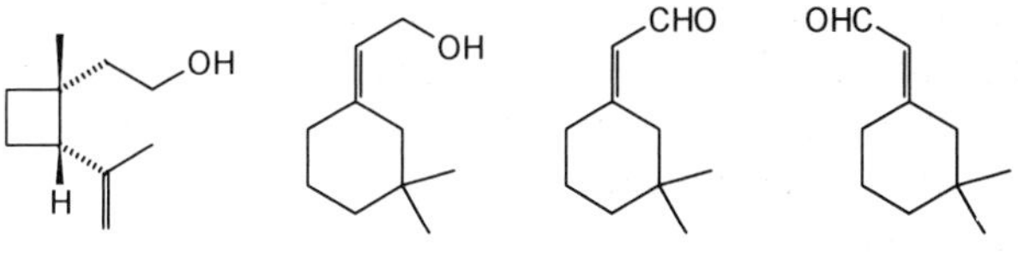

Fig. 7.1

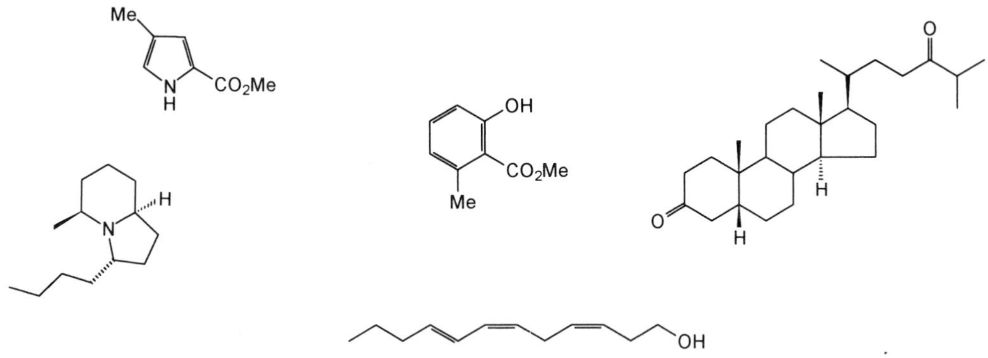

Fig. 7.2

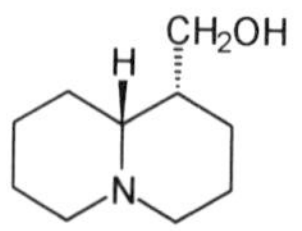

7.9

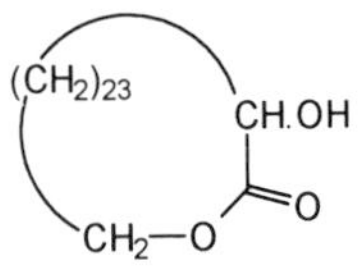

7.10

CH₃(CH₂)₁₁CH=CHNO₂

7.11

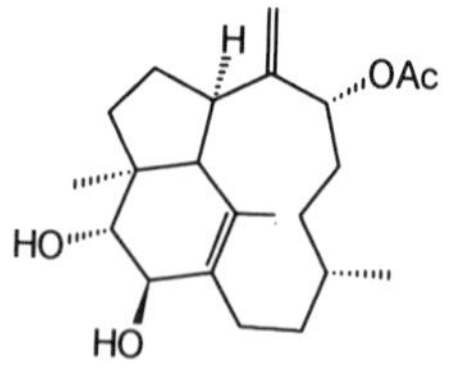

7.12

The trail pheromones employed by ants, termites, and certain insect larvae contain secondary metabolites produced by all of the main biosynthetic pathways, and some representative examples are shown in Fig. 7.2.

In addition to the production of pheromones, numerous insect species have developed means of storing toxic or distasteful secondary metabolites produced by plants. Storage of pyrrolizidine alkaloids is particularly widespread, and is especially prevalent in butterflies of the orders Danainae of North America and Ithomiinae of South America. These alkaloids are sequestered and used as a means of rendering the butterflies unpalatable, but also as an adjunct to mating. The alkaloids are transferred from the male to the female during courtship and copulation (as much as 1.5 mg), and then appear to act as a means of defence for the female and her offspring. The female, in consequence, appears to seek out those males that are rich in pyrrolizidine alkaloids since this will help to ensure her own survival and also that of her offspring.

Certain aphids also sequester quinolizidine alkaloids from their food plants, especially those found in members of the lupin family, e.g. lupinine, **7.9**. The aphids are thus spared the attentions of ladybirds and certain carnivorous beetles, though some species of ladybirds are able to feed upon the aphids and then store the alkaloids. This gives them additional protection, though they are also able to make their own deterrent compounds like coccinelline, **3.17**.

Insects also derive most of their actual offensive chemicals from their host plants, and termites release a veritable arsenal of injected poisons like **7.10**, contact poisons like **7.11**, and sticky secretions like **7.12**.

7.3 Plant–plant interactions

Many plants produce volatile chemicals (mainly monoterpenes) from their leaves which inhibit the respiration of other plants, or they release germination inhibitors (mainly phenols) from their roots. The compound juglone, **7.13**, from the walnut tree, gramine, **7.14**, from barley, and camphor, **7.15**,

7.13   **7.14**   **7.15**   **7.16**

and 1, 8-cineole, **7.16**, from the dominant plants of semiarid parts of southern California (chaparral) and the South of France (garrigue). Such inhibition of seed germination and plant growth due to the influence of secondary metabolites is known as allelopathy.

7.4 Stress metabolites

Most plants respond to the stress caused by microbial attack by the production of new secondary metabolites known as phytoalexins. Representative examples are shown in Fig. 7.3, and it can be seen that most biosynthetic pathways can provide phytoalexins. Many of these compounds have antiviral or antifungal activity.

7.5 Animal–animal interactions

Marine organisms produce a plethora of strange secondary metabolites, and most of these are apparently used for defence. Thus the pacific sole produces pavoninin I, **7.17**, which it uses as a shark repellant. This fish can release as much as 70 mg of a mixture containing this compound and other repellants, at the last moment of an attack, thus avoiding too much dilution by seawater. The limpet, *Collisella limatuala*, releases limatulone, **7.18**, which is a highly potent fish-feeding deterrent. The sponge *Smenospongia aurea* produces the bromoindole **7.19** which has antimicrobial activity, and the shark *Squalus acanthias* appears to employ the steroid **7.20** as an antifungal agent.

7.17   **7.18**

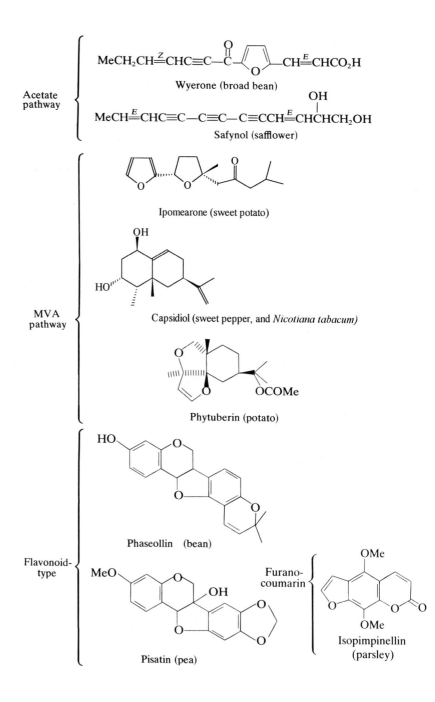

Acetate pathway

Wyerone (broad bean)

Safynol (safflower)

Ipomearone (sweet potato)

MVA pathway

Capsidiol (sweet pepper, and *Nicotiana tabacum*)

Phytuberin (potato)

Phaseollin   (bean)

Flavonoid-type

Furano-coumarin

Pisatin (pea)

Isopimpinellin (parsley)

Fig. 7.3

7.19

7.20

Numerous frog and salamander toxins have been identified, and batrachotoxin (see Fig. 1.1) and samandarine, **7.21**, are representative examples. These are usually stored in the tissues and released as secretions when the animal is attacked.

Attractive interactions are also known, and although mammals are not masters of 'perfumery' like the insects, many species do produce sex pheromones. Several well authenticated examples are shown in Fig. 7.4. There is some evidence that 5-α-androst-16-en-3-alpha-ol, **7.22**, is a 'spacing pheromone' for humans. The steroid is produced by both males and females, with the highest concentration in the latter just prior to ovulation. Females are most sensitive to this odour and there is some evidence that the synchronization of menstrual cycles observed in women's institutions (prisons, university halls of residence, etc.), is in some way induced by this steroid.

Clearly there is still much to be learnt about ecological chemistry, but the quest for a greater understanding of the complex interactions that occur between species provides a further rationale for the study of secondary metabolism. Human cultures have always exploited secondary metabolites as poisons, stimulants, narcotics, perfumes, and medicines. Thanks to the efforts of numerous investigators we now know how most classes of secondary metabolites are biosynthesized. However, we are still largely ignorant of why they are made and what factors influence the types of structures produced. Further studies at the genetic level (molecular biology) and in the wild (ecological chemistry) may provide the answers.

7.21

7.22

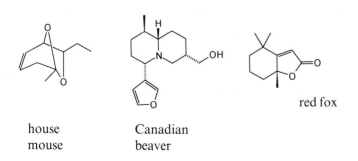

house
mouse

Canadian
beaver

red fox

Fig. 7.4

References and further reading

Generally applicable texts and journals with information on biosynthetic pathways.

1. Haslam, E. (ed.) (1979). *Comprehensive organic chemistry*. Vol. 5. Pergamon Press, Oxford.

2. Manitto, P. (1981). *Biosynthesis of natural products*. Ellis Horwood, Chichester.

3. Conn, E.E. (1981). *Secondary plant products*. Academic Press, New York.

4. Luckner, M. (1983). *Secondary metabolites in microorganisms, plants, and animals*. Second edition. Springer-Verlag, Berlin.

5. Torssell, K.B.G. (1997). *Natural Product Chemistry*, second edition, Apotekarsocieteten, Stockholm.

6. Haslam, E. (1985). *Metabolites and metabolism*. Oxford University Press, Oxford.

7. Mann, J. (1987). *Secondary metabolism*. Second edition. Oxford University Press, Oxford.

8. Herbert, R.B. (1989). *Biosynthesis of secondary metabolites*. Second edition. Chapman and Hall, London.

9. Smith, C.A. and Wood, E.J. (eds.), (1992). *Biosynthesis*. Chapman and Hall, London.

10. Dewick, P.M. (1997). *Medicinal Natural Products*. Wiley, Chichester.

11. Pattenden, G. (ed.) (vols 1–7) and Simpson, T.J. (ed.) vols 8 *et seq*. *Natural Product Reports*. Royal Society of Chemistry, Cambridge.

    This review journal was first published in 1984 and has appeared at the rate of six issues per year ever since. It is the best source of current information on all aspects of natural products.

Specific books and reviews for each chapter

Chapter 1

1. Staunton, J. (1978). *Primary metabolism*. Oxford University Press, Oxford.

2. Stryer, L. (1988). *Biochemistry*. Third edition. Freeman, San Francisco.

3. Lehninger, A.L. (1993). *Principles of biochemistry*. Third edition. Worth, New York.

Chapters 2 and 3

1. Rawlings, B.J. (1998). *Biosynthesis of fatty acids and related metabolites* in *Natural Product Reports*, **15**, 275.

2. O'Hagan, D. (1991). *Polyketides*. Ellis Horwood, Chichester.

3. Rawlings, B.J. (1997). *Biosynthesis of polyketides* in *Natural Product Reports*, **14**, 523.

Chapter 4

1. Harrison, D.M. (1990). *Biosynthesis of triterpenoids, steroids, and carotenoids* in *Natural Product Reports*, **7**, 459.

2. Beale, M.H. (1991). *Biosynthesis of C_5–C_{20} terpenoid compounds* in *Natural Product Reports*, **8**, 441.

3. Akhtar, M. and Wright, J.N. (1991). *A unified mechanistic view of oxidative reactions catalysed by P-450 and related Fe-containing enzymes* in *Natural Product Reports*, **8**, 527.

Chapter 5

1. Dewick, P.M. (1998). *Biosynthesis of shikimate metabolites* in *Natural Product Reports*, **15**, 17.

2. Haslam, E. (1993). *Shikimic acid: metabolism and metabolites.* Wiley, Chichester.

Chapter 6

1. Herbert, R.B. (1997). *Biosynthesis of plant alkaloids and nitrogenous microbial metabolites* in *Natural Product Reports*, **14**, 359.

2. Herbert, R.B. (1989). *Rodd's Chemistry of Carbon Compounds*, (S. Coffey, ed.), Second edition, vol IV, part L, p. 291, Elsevier, Amsterdam.

Chapter 7

1. Harborne, J.B. (1993). *Advances in chemical ecology* in *Natural Product Reports*, **10**, 327.

2. Harborne, J.B. (1988). *Introduction to ecological biochemistry.* Third edition. Academic Press, London.

Reviews on enzymological and molecular genetic advances.

1. Cane, D.E. (1991). *Tetrahedron*, **47**, pages 5919–6078.

2. Thiericke, R. and Rohr, J. (1993). *Natural Product Reports*, **10**, 265.

3. Cane, D.E. (ed.) (1997). *Chem. Reviews*, **97**, 2463–2705.

General texts on the use of natural products by human cultures: pharmacological and toxicological importance.

1. Lewis, W.H. and Elvin-Lewis, M.P.F. (1977). *Medical botany: plants affecting Man's health.* Wiley-Interscience, New York.

2. Mann, J. (1992). *Murder, magic, and medicine.* Oxford University Press, Oxford.

Index

LIVERPOOL
JOHN MOORES UNIVERSITY
AVRIL ROBARTS LRC
TITHEBARN STREET
LIVERPOOL L2 2ER
TEL. 0151 231 4022